THE NEW GEOGRAPHY
OF INNOVATION

THE NEW GEOGRAPHY OF INNOVATION

The Global Contest For Breakthrough Technologies

Mehran Gul

WILLIAM
COLLINS

William Collins
An imprint of HarperCollins*Publishers*
1 London Bridge Street
London SE1 9GF

WilliamCollinsBooks.com

HarperCollins*Publishers*
Macken House
39/40 Mayor Street Upper
Dublin 1
D01 C9W8, Ireland

First published in Great Britain in 2025 by William Collins

1

Copyright © Mehran Gul 2025

Mehran Gul asserts the moral right to be identified
as the author of this work in accordance with
the Copyright, Designs and Patents Act 1988

A catalogue record for this book is available from the British Library

ISBN 978-0-00-832780-4 (hardback)
ISBN 978-0-00-832781-1 (trade paperback)

All rights reserved. No part of this publication may be
reproduced, stored in a retrieval system, or transmitted, in any form or by
any means, electronic, mechanical, photocopying, recording or otherwise,
without the prior permission of the publishers.

Without limiting the author's and publisher's exclusive rights, any
unauthorised use of this publication to train generative artificial intelligence
(AI) technologies is expressly prohibited. HarperCollins also exercise their
rights under Article 4(3) of the Digital Single Market Directive 2019/790 and
expressly reserve this publication from the text and data mining exception.

Typeset in Adobe Garamond Pro by
Palimpsest Book Production Ltd, Falkirk, Stirlingshire

Printed and bound in the UK using 100% renewable electricity
at CPI Group (UK) Ltd

This book contains FSC™ certified paper and other controlled sources
to ensure responsible forest management.

For more information visit: www.harpercollins.co.uk/green

The present, accurately seized, foretells the future.
V. S. Naipaul

In loving memory of my mother, 1956–2015.

Contents

INTRODUCTION

Capitalism with an Emoji Face

'If you saw a bunch of people that were trying to do a thing and there were green people and red people and blue people and purple people, and the purple people are succeeding and the other ones are not, a natural thing to say is, well, what's up with the purple people?'

1

CHAPTER ONE

The Precocious Student

'America is the teacher, China is the student when it comes to great technologies; but China's now a precocious student who is learning from the teacher but also maybe out-executing the teacher, right?'

16

CHAPTER TWO

Steeples of Excellence

'I've learned not to bet against Silicon Valley because I already did that once and I was wrong.'

56

CHAPTER THREE

Busting Monasteries

'We're breaking it down from this being about the monasteries in Silicon Valley and in Boston and maybe one or two other places to a reformation, a renaissance back to Europe.'

98

CHAPTER FOUR

Hyper Gap

'Our mindset is we need to be so far ahead of our competition that it's a matter-of-fact thing to just give up because there's such a huge gap between the Korean companies and others.'

138

CHAPTER FIVE

Smart Nation

'If I'm asking myself as someone who's good with computers, what is the most effective pathway for me to make the world better, it would be to try and make governments better and provide technology tools for them to perform better.'

169

CHAPTER SIX

Small Wonder

'*Switzerland is a kind of trust, reputation is the brand.*'

203

CHAPTER SEVEN

The New Mittelstand

'*Silicon Valley is good for high tech,
but Germany is good for deep tech.*'

238

CHAPTER EIGHT

Importing Genius

'*It would be a shame if we invented this technology and then
had to buy the applications of it back from others.*'

270

CONCLUSION

Social Animals

'*Cultures are not like biological systems;
cultures are biological systems.*'

304

ACKNOWLEDGEMENTS

318

NOTES

321

INDEX

344

INTRODUCTION

Capitalism with an Emoji Face

'If you saw a bunch of people that were trying to do a thing and there were green people and red people and blue people and purple people, and the purple people are succeeding and the other ones are not, a natural thing to say is, well, what's up with the purple people?'

I

In the winter of 2000 when Mikko Kodisoja wanted to hire someone to help run his company, Sumea, a game developer based in Helsinki, Finland, he offered the job to Ilkka Paananen, a 22-year-old student at the Helsinki University of Technology. The most attractive aspect of Ilkka's candidacy was that he was the only one willing to take the job.

'These guys hadn't raised any money so they couldn't even afford to pay me anything,' Ilkka recalls. 'They didn't have too many candidates and I was probably the only one and that's why they picked me.'

Sumea's founders were creative types who were really into making games, but they didn't care much for all the other mundane stuff that goes into running a company. So they brought in Ilkka, a business student, to take care of all that housekeeping.

'And then they thought that, well, for you to have any kind of credibility as a kind of representative of the company you need a credible title and so they decided to call me the CEO,' Ilkka tells me. 'I'd never had a real job besides traineeships and some summer jobs and of course I had absolutely no idea what I was doing.'

Sumea would go on to be a moderately successful company. In 2004 it was bought by Trip Hawkins, an industry legend who practically kicked off the home gaming industry when he founded Electronic Arts in the 1980s.

After the acquisition Mikko and Ilkka paired up again to launch another gaming company where things could be exactly how they wanted them to be; 'the best people who would form the best teams which would then eventually create the best games'. They called the company Supercell.

Supercell, which started with fifteen people packed in a snug 35-square-metre room, got off to a rocky start. Its first game, *Gunshine*, didn't get much traction and was scrapped within months. The second, *Hay Day*, did well. The third, *Clash of Clans*, released in August 2012, changed mobile gaming forever.

Clash of Clans, a strategy game in which the player assumes the role of a village chief building out their own village while waging war on every other village, became the highest grossing game in the US within three months of its launch. Within a year it was the most profitable game in the world.

From idea to launch, *Clash of Clans* was developed in six months. A decade later it's still going strong. It's been downloaded over half a billion times and has brought in over $10 billion in lifetime revenue. How does that stack up against other hits coming out of the wider entertainment industry like movies, books and music?

The *Clash of Clans* franchise has made more money than the top three highest grossing movies ever, *Avatar*, *Avengers: Endgame* and *Avatar: Way of the Water*, combined. It has made more money than all the Harry Potter books, combined. It has made more money than the top hundred music singles streamed ever, combined.

In 2016, Supercell was acquired by the Chinese tech giant Tencent in a deal that valued the company at over $10 billion, making it the first European tech company of the internet era to cross an eleven-digit valuation. Supercell employs fewer than 400 people, making it the most valuable company per employee in the world. The Finnish company's runaway success has turned Ilkka, the student who became CEO because no one else wanted the job, into a billionaire.

Supercell was started in part with a €400,000 loan from the Finnish government without which, Ilkka says, the company wouldn't exist. That investment has paid off massively. The Helsinki-based company is the single largest corporate taxpayer in Finland, the biggest global success to come from the frosty reaches of this tiny Nordic country since Nokia. Supercell alone pays back multiples of all the money invested by the Finnish government in every startup ever.

Supercell is as much a product of its environment as it is a creation of its founders. If it takes a village to raise a child, it takes a city to raise a company. Without good schools, quality healthcare, safe streets and strong social security nets, Ilkka says, a company like his would not exist. 'We are proud of our Finnish roots,' he tells me. 'We feel that in Finland we have a social system and a community which has enabled us to be successful.'

2

Ajay Agarwal spent most of his career as a management professor thinking about one question. A lot of places are good at science. But few can turn that science into new technologies and big companies. Why?

This disparity made Agarwal, who is the Geoffrey Taber Chair in Entrepreneurship and Innovation at the University of Toronto's Rotman School of Management, study how new science is turned into commercial products and viable companies in different places.

'It became a geography question,' he tells me. 'If you saw a bunch of people that were trying to do a thing and there were green people and red people and blue people and purple people, and the purple people are succeeding and the other ones are not, a natural thing to say is, well, what's up with the purple people?'

The answer wasn't to be found in obstruse academic debates. 'At some point I decided, no more round tables, no more white papers, no more breakfast lunches and all these things,' he says. 'We're just going to try and build a thing, even if it's very small, and that's what led to the Creative Destruction Lab.'

The Creative Destruction Lab, or CDL, is a not-for-profit incubator of science-based companies launched at the University of Toronto in 2012 with the goal of 'enhancing the commercialization of science for the betterment of humankind'. It runs a nine-month programme to help promising founders get their early stage, science-based ventures off the ground.

When CDL first launched in 2012, Agarwal struggled to find people to mentor his founders. There just wasn't a critical mass of people who had built a successful tech company out of Toronto. So he decided to bring in tech heavyweights from the US every couple of months to simulate Silicon Valley in Toronto for moments in time.

'We had to jump-start it the way you jump-start a car: you've got to somehow get the spark going,' he says. 'The way to understand the Creative Destruction Lab is it temporarily creates a Silicon Valley in Toronto by bringing together all of the actors that make the Silicon Valley thing work.'

Within a relatively short time CDL has graduated companies pursuing wildly ambitious goals using borderline fictitious technologies that only a decade ago, it would've been thought, could only come from Silicon Valley or Boston or Seattle.

These include Xanadu, a company that builds quantum computers using light particles; Deep Genomics, which uses machine learning to decode the molecular basis for genetic diseases; and Kepler, which is developing an interstellar communications network to extend connectivity to deep space. All three are based in Toronto.

'The space stream at CDL attempts to create an ecosystem where someone with a cool idea can actually see that there is an incubator here in Canada that is going to put together the business acumen and the financial resources to be able to turn it into a business,' Chris Hadfield, a retired Canadian astronaut who has served as the commander of the International Space Station, tells me. Hadfield now heads CDL's space stream.

Agarwal thinks that CDL has only just kicked off a virtuous cycle where experience and wealth from successful companies gets reinvested into the ecosystem to create more successful companies: 'It's like a flower that releases its pollen,' he says.

And while CDL's initial spark might have come from the ecosystem he imported from the Valley, Agarwal pushes back on my suggestion that he is engaged in some kind of social experiment to clone Valley culture in Toronto.

The Valley has many virtues but also many vices, he argues, and it is not his intention to replicate the 'bro culture' or 'the aggressive, hard-driving culture' that he sees in the Valley. 'I would definitely not say we are trying to replicate Silicon Valley culture,' he says. 'I would say a hundred per cent not.'

3

When the French billionaire Xavier Niel wanted to build the world's largest startup superhub in Paris so that the city could compete with the likes of the Bay Area, Boston and London in tech, he made the unusual choice of picking a 30-year-old American tech journalist who majored in French literature at UCLA to spearhead its development. 'I thought Xavier was pulling my leg,' says Roxanne Varza, Director of Station F, the world's largest startup facility. 'I even recommended someone else for the job.'

If there is a beating heart to the Parisian tech scene it is Station F. The campus, housed in a repurposed 100-year-old train depot that spans the size of four football fields, is designed to bring an entire entrepreneurial ecosystem under one roof. Its founder, Xavier, poured €250 million of his own money into building this giant sandbox where he could assemble a superdense concentration of all the various talents that are needed to build supermassive companies. The building has space for a thousand new ventures.

Though an entirely private initiative, Station F has become something of a symbol for France's ambition to become the tech capital of Europe. When Emmanuel Macron inaugurated the facility in 2017, he made a pitch for why this is the place for the smartest people with the biggest dreams. 'I like to compare a researcher in Harvard with a researcher in France,' he said. '[In France], school is free and excellent, healthcare is free, there's a retirement system. On the other side, there's nothing.'

Station F is clear about its goals: it wants to be the destination of choice for talented founders from all over who want to build the next Apples and Googles of the world. 'We needed to have a successful unicorn company come out of Station F,' says Roxanne. 'I feel like we had to tick that box, otherwise we're not successful.'

It took them five years to pull that off. In 2022, Hugging Face became the first Station F alum to cross a billion-dollar valuation.

The young company took almost no time to become the central hub of the AI community. It's the most widely used online platform where AI experts and enthusiasts come together to share what they are working on and collaborate on building machine learning models.

Hugging Face is named after the smiley face with jazz hands emoji. Its founder, Clement Delangue, has said that when the company goes public he wants it to be listed with an emoji instead of a three-letter ticker. Capitalism with a human face, out; capitalism with an emoji face, in.

For Roxanne, Hugging Face's breakout success validates Station F's founding hypothesis that big things can come from France. It might not be long before the flow of talent is from Palo Alto to Paris and not the other way around. After all, that's the choice she made herself.

'I almost think it's more interesting as an individual to not be in the Bay Area if you want to do what I'm doing,' she says. 'From the moment I arrived in France, I've been able to see my direct impact, so I think for what I'm doing, this is a more interesting place to be.'

4

The US is the source of just about all the technologies that define modern life: personal computers, operating systems, smartphones, e-commerce, web browsers, email, search engines, social networks, electric cars and the rest. And most of the tech companies that created and monetized these technologies are also in the US. This book asks: is that changing?

This question comes up most often in the context of China. Less than a decade ago, the sentiment towards Chinese tech companies was often dismissive and complacent. Now the alarm bells are ringing. China is no longer just about a handful of old tricks like cashless payments or WeChat or 5G. Its competence in tech is broad and deep. ByteDance, the creator of TikTok, is the most valuable privately held company in the world. BYD makes more electric cars than Tesla. DJI sells more commercial drones than everyone else combined.

But as the commentariat pontificates how the US–China tech battle will play out, an equally interesting question to ask is: are there more Chinas out there? Places no one is taking seriously now that might turn out to be massively competitive sooner than we think.

The US and China monopolize most of the airtime, and just given their sheer size they are probably the two tech powers that matter the most, but that doesn't mean that they're the only two that matter. At least a dozen or so countries are producing globally relevant companies that make things that touch the lives of billions of people around the world every day. And some of these places, when adjusted for population size, are even more prolific at producing new tech than the two big tech rivals.

Samsung, a South Korean conglomerate, competes with Apple to be the world's largest manufacturer of smartphones. One in four people who use a smartphone use a Samsung. Arm, founded in the UK, develops chip designs that are used in more than 90 per cent of all mobile devices. Spotify, based in Sweden, is the most popular music streaming service in the world.

That's not all. The world's most important semiconductor company, TSMC, is in Taiwan. The other most important company in the semiconductor industry, ASML, is in the Netherlands. Some of the world's best-known games like *Minecraft*, *Candy Crush* and *Angry Birds* came from gaming studios in the Nordics. Nearly all the major electric battery manufacturers like CATL, LG, and SK On are in Asia.

This is a story about technology and the places where it finds its way into the world. Silicon Valley has for half a century been unrivalled in spinning out technologies and fast-growing, high-value, billion-dollar-plus tech companies, the Apples, Facebooks, Googles of the world, that made it the centre for the most rapid creation of wealth in human history. Its secrets are spreading to more places.

The geography of innovation is shifting. The world has a lot more high-value tech companies than ever before, growing a lot faster than ever before, in a lot more places than ever before. This is a book about these places.

5

So what are these places? It depends on how you look at it. I use three different approaches in this book.

The first is to look at things from the perspective of a venture capitalist. Here the measure is simple: who's producing the most billion-dollar startups? It's a narrow but useful, quick and dirty filter for relevance. Seen through this lens the map of the most innovative places in the world looks like this.

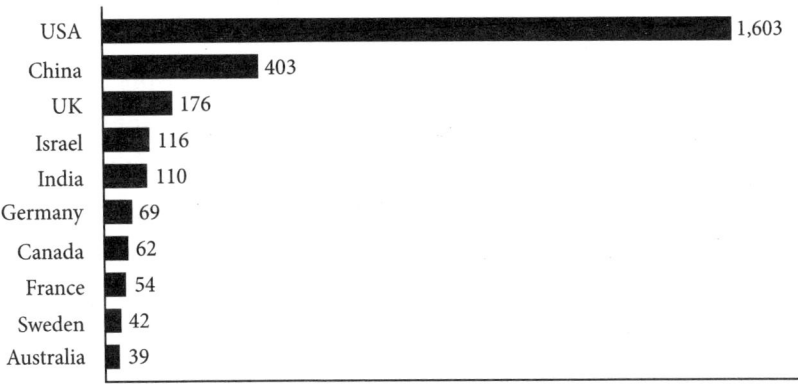

Figure 1: Top ten countries by number of private billion-dollar tech companies

Source: Dealroom (2025)

A helpful complement to counting high-value tech companies would be to add another layer to the analysis: who's attracting the most venture funding? This produces a slightly different list.

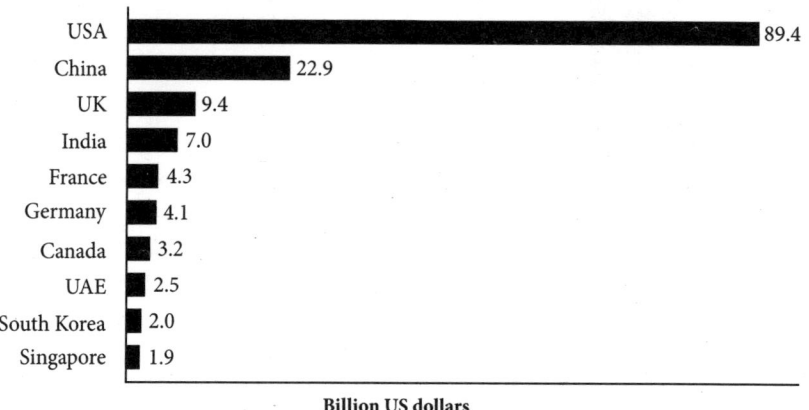

Figure 2: Top ten countries by volume of venture dollars invested

Billion US dollars
Source: Dealroom (2025)

Filtering by the number of high-value startups and the amount of funding they attract is a useful way of looking at things, but not without drawbacks. The most obvious being that it assumes that innovation only happens in new companies. Nvidia, founded in 1993, is now comfortably in its midlife. And yet arguably it's only just hitting its stride.

And here's where the second approach comes in. This considers both old companies and new ones. Here we measure how big a region is in tech by looking at the cumulative market cap of all tech companies based there. Seen this way the world looks something like this:

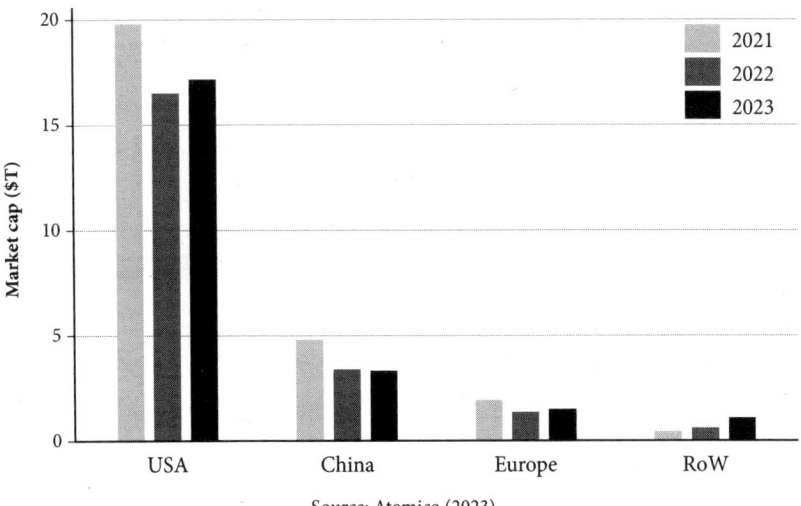

Figure 3: Total tech market cap ($T) per region, 2021–2023

Source: Atomico (2023)

There are many ways to determine the rank order of the most innovative places. But I think the graph above most accurately captures the relative importance of different regions to the magnitude of technological change that we see in the world today. This relatively narrow filter of tech valuations maps neatly to where the most amount of tech innovation, understood in the broadest possible sense of the term, is happening globally. That's my feel for what the picture looks like developed over the course of writing this book. We'll discuss why in the following pages.

Both of the above approaches are open to criticism, chiefly because they assume that tech innovation is the only kind of innovation that matters and even that is being measured with narrow financial metrics: valuations and venture dollars. Surely there's more to it than that. Fair enough. Let's bring in a third lens, the Global Innovation Index, or GII, a ranking of the most innovative countries in the world published by the World Intellectual

Property Organization, or WIPO, a UN agency based in Geneva, Switzerland, which counts 193 nation states among its members.

Their approach is anything but reductive. WIPO claims to provide a comprehensive assessment of the state of innovation in the world by crunching over eighty indicators which range as widely as research output, R&D expenditure, education spending, test scores, valuations, engineering graduates, patents and so on. According to this impressively multifactorial analysis, the rank order of the most innovative places in the world looks something like this:

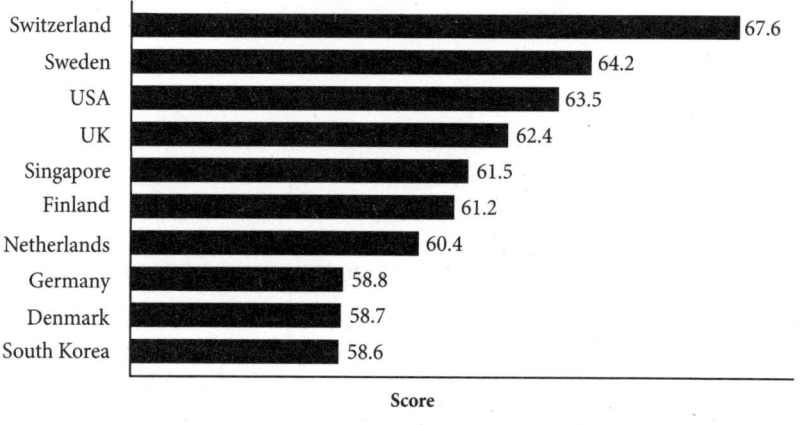

Figure 4: Highest ranking countries in the WIPO Global Innovation Index

Source: WIPO (2023)

That's a fine list. But not one above reproach. The GII's heavily quantitative bent lends this benchmarking exercise a patina of legitimacy grounded in mathematical rigour but arrives at conclusions that don't quite pass the smell test.

Seven out of ten countries on that list are European, giving the unmistakable impression that the continent is firmly at the centre of where the future is being made. That does not ring true. And European policymakers would be the first to point that out.

'The EU is losing the race for innovation,' writes Nobel Prize winning economist Jean Tirole in a recent report by the European Policy Analysis Group. The document notes that the transatlantic gap in new technologies is widening and China too is at the cusp of leaving the continent behind.

Speaking of which, China, the only country that's been giving the US a run for its money in the race to develop new technologies, is also conspicuous by its absence from that list. What gives?

Reminds me of an old quote from Keynes: 'too large a proportion of recent "mathematical" economics are mere concoctions, as imprecise as the initial assumptions they rest on, which allow the author to lose sight of the complexities and interdependencies of the real world in a maze of pretentious and unhelpful symbols.' Sometimes when one unpacks all that mathematical rigour, all one finds is a kind of intellectual rigor mortis.

But even if we can quibble with how the GII came up with those numbers, the basic idea, that innovation is about more than just tech companies and their valuations, is a laudable one, and we're going to hold on to that.

Each of these three approaches has its limitations. But taken together they present three useful entry points for an enquiry into which places matter the most in bringing newness into the world. That's a lot of countries so, to start, we'll mix and match. This first book is weighted towards North America, Europe and the Far East, which is where the bulk of the action is. In subsequent books we'll broaden the discussion to more places.

This book is organized geographically. Each chapter provides the reader with an in-depth look into what's happening in a specific country. But there are two threads that cut through these discussions. The first: what makes a place innovative? Regardless of which filter we use to sort and rank the winners, the fact is that there are precious few of them around. Why?

Most discussions about innovation revolve around personalities and institutions. Steve Jobs was a creative genius. Pixar built a unique culture. Stanford redefined what it means to be a university. That sort of thing. But this book goes one level higher to look at the wider ecosystem from which these people and institutions and companies emerge.

The view that it takes is that companies are a bit like seeds that are planted in their environment. New technologies are the fruits of these seeds. And bad farmers grow weeds, good farmers grow crops, and great farmers grow soil. So what is it about the entrepreneurial soil of these environments that makes them such fertile grounds for new ventures?

The second thread that runs through this book is: what does it even mean for a place to be innovative? Is it just about having companies like Apple, DeepMind and ByteDance? Or is there more to it? And what, exactly, is that 'more'?

To be sure, this book leans heavily towards a survey of the tech company type of innovation. Three reasons. First, Occam's razor: it keeps things simple and focused. The discussion must occur within some sort of bounds otherwise it will veer into all sorts of unhelpful directions.

Second, because it does seem that the most consequential developments in technology are no longer happening in universities or government labs or corporates but within the context of fast-growing startups. So if one had to focus on that one thing that is most driving tech innovation, it would make sense to focus on that.

And third, and this is the controversial one, arguably all the other measures of innovation – all that R&D, government spending, STEM graduates, patents, research output, university rankings, venture dollars – the score from all those inputs should tally up to new technologies. That's the river that all the tributaries flow to, so it makes sense to skip the intermediaries and take stock of that end product. And that's mostly happening in companies. So that's the focus. But we do at times pause and ask, is there a better way of doing this?

This is a book about technology. But it has people at its centre. Specifically, the people who are making that technology happen. It is the product of conversations with almost 200 prominent figures in technology around the world: entrepreneurs, scientists, venture capitalists and public officials who have made it their vocation to turn the wheel of progress. It's an attempt at piecing together a coherent picture of what the map of the most technologically capable places in the world looks like, how they got that way, and the world they are making. At the most basic level it asks: what's going on out there? It's a question that turned out to be a lot harder to answer than I had expected.

CHAPTER ONE

The Precocious Student

> *'America is the teacher, China is the student when it comes to great technologies; but China's now a precocious student who is learning from the teacher but also maybe out-executing the teacher, right?'*

Expansion

In the summer of 2000, David Wallerstein, a fresh-faced 25-year-old consultant at Naspers, a South African conglomerate, flew out to Shenzhen to meet Ma Huateng, then a 29-year-old co-founder of a Chinese internet startup that ran an online chat service called QQ, the Chinese variant of ICQ, AOL's PC-based messaging service. QQ, released in 1999, had been hugely popular with China's newly connected youth and grew to a million users within a year of its launch at a time when the entire country had only about 22 million people connected to the internet. Wallerstein was sitting on a cash pile of $100 million, a huge sum at the time, that Naspers had put aside to invest in China's nascent internet economy. He was making his way to Shenzhen with what he thought was a pretty convincing offer to buy Huateng's young company.

Wallerstein was used to having good meetings in China. The country was still poor, the average person made less than $100 a month, foreign investment was scarce, and the very idea of a technology startup was little more than a poorly understood novelty. Wallerstein, and his $100-million chequebook, were usually a welcome sight wherever they went. But not this time. After exchanging pleasantries, and a brief meeting, Ma Huateng, also known as Pony Ma, politely told Wallerstein that his two-year-old company, Tencent, was not for sale. Thank you for coming and goodbye.

Wallerstein was taken aback. He had never really had this experience in quite this way before. But he hadn't come all the way from Beijing only to be told to go back to where he came from. Undeterred by the snub, he invited Huateng and his four co-founders to dinner that night where, as legend goes, everyone got plastered. The next morning, he went back to Tencent's office to make another pitch for how the two companies could work together. It took over a year of courtship before Huateng finally came around and sold half of Tencent to Naspers for $34 million, a bold investment for that time and place, eclipsing as it did even Softbank's much mythologized $20-million bet on Alibaba that happened only a year earlier.

After the investment, Wallerstein, a native Californian who attended the University of Washington and Berkeley, and who likes to play rock guitar in his spare time, was hired by Tencent as the company's first non-Chinese executive. He became the first person outside of the company's five founders to join Tencent's executive team, the firm's highest level of leadership. At the time, the three-year-old company had forty-five employees and one product, QQ, which had only just managed to turn a profit. 'In the early days I used to do speeches to try to rally the troops, to encourage the company that we could become an international brand someday, that we could be a participant in the global economy,' Wallerstein, who prefers to work out of a converted church building in Palo Alto, tells me. 'These were very hard to imagine concepts in 2000.

I think even after we went public, it still wasn't really clear to us that we could have any role or impact in international markets.'

At fifty, Wallerstein has devoted fully half his life to Tencent and was for the longest time the highest ranking and still is the longest tenured American in the upper ranks of the business. He is among a handful of Westerners swept along with the improbable rise of Tencent, and China, from near oblivion to almost total relevance in all things technology in the impossibly compressed time span of just twenty-five years. At its peak, in early 2021, Tencent was, with a market cap scraping almost $1 trillion, the most valuable company in Asia. Naspers' $34 million early bet grew to be worth over $200 billion, making it one of the most outrageously successful investments in the history of capitalism, the corporate equivalent of winning the Powerball.

Naspers started as a modest newspaper business in Stellenbosch, South Africa, in 1915 and spent much of the twentieth century publishing Afrikaans-language newspapers and magazines. This single investment in an obscure Chinese startup initiated by a 25-year-old consultant transformed it into a global internet and media powerhouse, at one point breaking into the ranks of the hundred most valuable companies on the planet, punching at the same level as tech headliners like Amazon, Netflix and Facebook. Naspers' stake in Tencent, which accounts for just about the entire value of the company, has made it by quite a wide margin the most valuable company on the entire continent of Africa.

How did Tencent grow so big so fast? In its early years the company's growth was fuelled almost entirely by its desktop-based messaging app, QQ. The company's first product, still used by 800 million people, made it worth mere billions, but not yet hundreds of billions; China famous, not world famous. The company's stratospheric rise would come more than a decade after its founding when it switched its focus from desktop to mobile devices. In 2011 it launched WeChat, the smartphone-focused superapp which would take Tencent, and the Chinese internet experience, to a whole new level. WeChat combines the functionality of Facebook, WhatsApp, Uber, Instagram and a variety of other payment and retail features; everything anyone would want to do on the internet on a smartphone all in one app. With WeChat, Tencent controls the single most important digital service in the world's single largest digital market. Nearly everyone – and I mean everyone – in China uses it. So central is WeChat to daily life in the country that a smartphone stripped of Tencent's flagship product would to an average user be rendered practically useless.

Wallerstein's first title at the company was Head of International. He spoke fluent English and Mandarin, had his formative experiences in the US, Asia and Africa, and split his time between the West Coast and China. So it was only natural for his role to be the bridge between Tencent and the outside world. The job of most executives with this responsibility in most companies is to take what they're selling at home and try and sell it abroad. Wallerstein couldn't do that. 'Everything was so optimized for China that our services didn't look attractive to anyone outside of China,' he recalls. 'So we actually didn't really focus on doing much with our products overseas.'

Wallerstein's role was less to bring Tencent to the world as it was to bring the rest of the world to Tencent. In the early 2000s, internet usage in China was beginning to take off and the country was adding tens of millions of new internet users every month, many of whom were curious to figure out what they could do with this newly acquired window into the outside world. Tencent saw an opportunity in catering to the newly connected Chinese consumers' ravenous appetite for foreign music, foreign games, foreign everything. It was Wallerstein's job to sate this demand. 'I was much more focused on bringing great technology, great assets, great intellectual property to China,' he says. 'And the core of that strategy was to license great games to come into China.'

Tencent's first forays outside the mainland started simple enough. Wallerstein partnered with foreign game developers to distribute their products on Tencent platforms in China. If the games did well, they split the profit. But over time the relationships grew deeper. Tencent went from distributing foreign games in China to buying stakes in the overseas companies that made these games. 'We found that a lot of these studios that we were interested in didn't have much capital and we didn't know if they were going to survive,' says Wallerstein. 'But we liked their games. So how do we make sure they can survive? Well, maybe we'll also invest in them and own a little bit of them to make sure they've got capital to finish their game.'

QQ was Tencent version one; WeChat, version two. The company's modest and entirely opportunistic early investments in small gaming companies paved the way for Tencent version three: its metamorphosis from a China-focused smartphone app developer to a highly diversified global technology conglomerate, sometimes called the Berkshire of tech.

Tencent's initial attempts to enter the gaming industry, which was hitherto dominated by American and Japanese players, were for the better part of half a decade met with spectacular failure. Those growing pains are now a distant memory as the firm has matured into the largest game vendor in the world. Tencent has stakes in over a hundred gaming studios from the US to Europe to Japan and South Korea and counts in its portfolio some of the industry's most coveted titles. It owns all of Los Angeles-based Riot Games, the maker of *League of Legends*, which attracts over 180 million gamers a month; almost half of Epic Games, the North Carolina-based creator of *Fortnite*, played by 350 million people; and over 80 per cent of Supercell, the Finnish gaming studio behind the hit *Clash of Clans* franchise which attracts over 100 million users a month.

Tencent has minority stakes in Activision Blizzard, the American gaming studio behind *Call of Duty*, *World of Warcraft* and *Candy*

Crush; the French video game publisher Ubisoft that created the blockbuster *Assassin's Creed*; and Krafton, the South Korean publisher of *PUBG*. It's hard to think of a popular game that is in some way *not* associated with Tencent. Grinding Gear Games (New Zealand), Funcom (Norway), Sharkmob (Sweden), Sumo Group (United Kingdom), Inflexion (Canada), are all Tencent properties.

Tencent the social media company attracts over a billion users. But what is less well known is that Tencent the gaming company also reaches over a billion people. The difference is that while the user base for the firm's social media apps is almost entirely Chinese, the audience for games linked to Tencent spans the globe.

Tencent's presence in the entertainment industry goes beyond gaming. It is also a major player in the global music industry. Tencent Music Entertainment, or TME, is by far the most popular music streaming service in China. It would be tempting to call it the Spotify of China, but that would be to make the parallel the wrong way around. Not only does Tencent's music streaming service have more users than Spotify, the most popular music streaming company outside China, it is in fact Spotify's third largest shareholder, so the relationship can hardly be termed competitive. It is also one of the largest shareholders in Universal, the largest music company in the world. Tencent is also big in movies. Its production arm, Tencent Pictures, has produced major American franchise movies like *Terminator: Dark Fate*, *Men in Black: International* and blockbusters like *Kong: Skull Island*, *Wonder Woman* and *Venom*.

When most people outside China think Tencent they still think WeChat. But that mental model is outdated. The company's rapid overseas expansion has taken it beyond social media, beyond gaming, beyond entertainment, beyond China, beyond categorization really, and turned it into a highly diversified global holding company with stakes in over a thousand companies, more than 600 of which are outside China, including prominent US tech brands like Snap, Reddit and Discord. In 2017, it paid $1.78 billion to buy 5 per cent of Tesla, making it the tech icon's fifth largest shareholder at the time. In 2021, at the peak of its international M&A activity, Tencent had amassed a listed investment portfolio of $190 billion, with stakes worth tens of billions of dollars more in unlisted companies.

In 2014, David Wallerstein transitioned into the role of the Chief eXploration Officer, or CXO, which made him the de facto head of Tencent's internal venture capital fund. His remit was to look for bleeding edge technologies wherever they could be found; radical moonshots that could make the company even more money while also plausibly solving some of humanity's biggest challenges, like health, energy and sustainability. From his office in Palo Alto he oversaw investments to the tune of billions of dollars into some pretty far out ideas, like Moon Express, a startup that aims to put drones on the lunar body, and Planetary Resources, a company that is looking into asteroid-mining.

Tencent was now big enough to be able to splash around 'utopian money', remaking it in the image of its peers in Silicon Valley whose business plans at times sound like they've been plagiarized from the pages of sci fi novels; less about monetizing captive markets or maximizing shareholder value and more about solving AGI, making humanity a multiplanetary species, and extending the light of consciousness. Wallerstein's job was no longer just about doing well. It was about doing good. When he came out with his first rock album it included the song 'The Last Chance', in which he sings a hook that goes: 'This is a chance, to face the reality/ One last dance, for all of humanity,' and ends by fading out to the sound of a ticking clock.

Wallerstein started his career bringing outside capital to promising companies in China. Over time his job became precisely the opposite: to invest Chinese capital in companies abroad. The arc of his career is a micro expression of one of the biggest trends unfolding at the level of the global economy. For decades China was the largest recipient of foreign direct investment in developing countries. Now more foreign investment flows out of China than into China. In 2023, inbound foreign investment into China totalled $33 billion, the lowest in three decades. In 2022, outbound foreign investment from China totalled $163 billion, the highest ever.

It's not just David Wallerstein, it's not just Tencent, and it's not just the tech industry. Chinese private and state-owned enterprises have been acquiring foreign assets across the board, encouraged in part by the government's 'going out' strategy. China is today the world's second largest contributor to global outbound investment.

The deal-making spans industries and geographies. Anbang, a major Chinese insurance company, owns the Waldorf Astoria in New York. Geely, a Chinese car maker, owns Volvo, the Swedish icon. The BAIC Group, a state-owned Chinese car company, is the largest individual shareholder of Mercedes-Benz. Sinochem, a Chinese state-owned chemicals company, owns Syngenta, the world's largest agriculture chemicals company, based in Switzerland.

The deal-making frenzy has not gone unnoticed. The US has been cracking down on Chinese equipment manufacturers like Huawei and ZTE for quite some time. It is now extending these restrictions to Chinese investors as well. In January 2024, the US Department of Defense added International Data Group (IDG), one of the more prominent Beijing-based venture capital funds, to a list of entities identified as 'Chinese military companies' supporting the 'modernization goals' of the Chinese military by 'ensuring it can acquire advanced technologies and expertise'. IDG, which started out as an American venture firm based out of Boston that was later bought out by its Chinese team, denies these allegations. This is the first instance of a major Chinese-headquartered investment firm being added to a sanctions list by the US government. The charge carries added weight given it comes not from a civilian agency tasked with overseeing trade relations like the Department of Commerce but from national security circles at the Pentagon. The DoD would eventually drop IDG from its entities list but the episode highlights the heightened geopolitical risks for investment firms operating across the US and China.

US companies were until only a few years ago liberally taking money from Chinese investors. That has come to a halt. 'There was a time in 2017–2018 where companies that were extremely capital intensive could look to China as a source of capital and they obviously can't anymore,' Shahin Farshchi, General Partner at Lux Capital, a US venture firm, told me. 'Now is that necessarily a bad thing? I don't think so. I think companies should be better at managing their cash and I feel like there's enough capital in the US. So if a company has a promising enough business case, then they should be able to raise that capital in the US.'

US regulators have also been curtailing capital flow the other way, from the US to China. In September 2023, President Biden signed an executive order banning private US funds from investing in certain technologies in China and requiring prior authorization for others. Several prominent US VC firms like Sequoia Capital, GGV Capital and DCM Ventures have either formally split from their China operations or indicated plans to reduce or restructure their China investments. Keith Rabois, Managing Director of Khosla Ventures, a major US venture firm, summed up the sentiment when he told me: 'We think China is an authoritarian regime that is ideologically bankrupt and morally suspect. So we don't want to participate in accelerating the power of the CCP.'

It was not long ago that the US was the largest single source of outside capital in Chinese tech startups and China was the largest single source of foreign investment in US tech firms. That's changed. This decoupling comes at a time when venture firms in the two countries are facing more saturated domestic markets, driving them to look for more opportunities abroad. 'I would be willing to fund companies basically anywhere around the globe,' says Keith Rabois. 'We don't have a hypothesis about where they're located. We will fund them. They can be anywhere – except China.' As the world's two largest economies reduce their exposure to each other, much of the capital that was previously being exchanged between them is now being redirected elsewhere, creating new opportunities for companies based in geographies previously overlooked by big tech investors.

One of these beneficiaries is Europe. The continent has long been seen as a distant third when it comes to the strength and competitiveness of its tech sector. While the largest companies in the US and China are tech icons like Apple and Tencent, in Europe the corporate landscape is still dominated by older firms from traditional sectors like luxury and pharma. One reason the continent hasn't produced globally relevant tech companies is because European entrepreneurs just don't have access to the massive amounts of risk capital needed to build them. Chinese investors are now stepping in to fill that void.

Alibaba and Didi Chuxing have invested hundreds of millions into European fintech startups like N26, Lydia and WorldFirst. Kuka, one of Germany's largest robotics companies, was acquired by the Midea Group in China in 2022. Tencent has stakes in dozens of European gaming studios, including Ubisoft, the continent's largest.

Like the US, European policymakers have also been stepping up scrutiny of Chinese investments in strategic technologies. The German government blocked the acquisition of Aixtron, a semiconductor manufacturer, by China's Fujian Grand Chip Investment Fund, reportedly at the urging of the US government. NEURA Robotics, a German company that makes robots that can perceive their environment and work with humans, also known as cognitive robots, or cobots, was founded with an $80 million investment by a single backer: the Han's Group, a Chinese conglomerate. In 2023, it had to buy out its Chinese backers and shift its entire production from China to Germany in view of, in the words of the company's founder, 'today's deglobalized world'. An investor in the company told me that it had brighter prospects now that it could move forward with 'a clean cap table'.

It's not just about security. It's also about reciprocity. 'So the challenge with China is it's very hard for a non-Chinese player to win inside of China, the game is very rigged and it's not fair to outsiders,' says Hussein Kanji, a venture capitalist based in London. 'We can't play there but they can play here.'

In the US, the outlook on China is consistently negative. In Europe, the feeling is more ambivalent. Strong suspicions of Chinese motives appear alongside what is at times a robust defence of closer economic ties with the world's second largest economy and the EU's largest trading partner. European leaders have been urging more scrutiny of deals involving China while also bristling at suggestions that decisions about who Europe can and cannot do business with can be made anywhere other than Brussels. European Commission President Ursula von der Leyen has emphasized the need for Europe to maintain its 'strategic autonomy'.

A senior European official told me that when they think about technology competition they are wary of rivals to the east and the west. 'It's not about values, it's about power,' they said. 'And it's very striking that people realize that the last two platform shifts of internet and mobile created just a staggering amount of US soft and sometimes hard power. And I think there is a general recognition, certainly Macron gets it, and Starmer gets it, that AI looks to be a platform shift or technological shift on the same scale. And if Europe doesn't try and play, then it will just exacerbate what we saw in the kind of post 1997/98 era.'

China is sometimes seen as the antagonist of European industry: the low-cost rival, the black hole of intellectual property. But it is also at times its unlikely saviour. Lilium is a Munich-based business which makes electric flying cars. The company, founded in 2015, is an ambitious oddity in an otherwise staid startup environment. It went through its worst crisis in early 2023 when it sustained losses of over $390 million, tanking its share price to below a dollar. Faced with an imminent delisting from the Nasdaq and unable to raise the required capital from domestic investors, the company turned to Tencent, which effectively saved it from certain death with a cash infusion of $175 million. David Wallerstein sits on the board of Lilium.

'It was a lifesaver for Lilium, it was wonderful what they did there,' says Hermann Hauser, the British-Austrian co-founder of Arm, one of the most prominent tech companies to come out of Europe. 'We need to do deals with China. It's just stupid. You cannot ignore the biggest market, the biggest economy in the world. People are so dollar focused and still so US centric that few people know that China is the largest economy in purchasing power and soon will be in dollar terms as well. They are already the lead market in a number of areas like smartphones and payment systems and electric vehicles. BYD has just overtaken Tesla as the largest EV manufacturer in the world. So it's just stupid to ignore this. You've got to deal with China whether you like it or not.'

Hauser, who now heads Amadeus, a Cambridge-based venture fund that focuses primarily on deep tech startups in Europe, spends much of his time advocating for technological sovereignty, the idea that Europe needs to stake out a more independent position on technological matters. 'I think at the heart of Europe there is a very strong feeling that Europe does have to be independent of the US *and* China,' he says. 'We certainly have the ability to be independent . . . there's a lot of pressure from the US being exerted on Europe, especially on the China policy, but I'm not politician enough to know if that is in the same way that the Americans are very vociferous about decoupling from China and saying we're not going to deal with China, and then if you look at what actually happens, they're increasing their trade by 25 per cent since they started saying that. European politicians will also say the right things to keep the Americans happy, but what will they actually do? You have to look at the trade figures, and I think the trade figures are still quite encouraging.'

European entrepreneurs are feeling the heat from US–China tensions. They need large sums of foreign capital to build companies at scale which sometimes can only come from countries like China but are wary of getting caught in the crossfire of a rivalry in which they often see themselves as neutral bystanders. Tencent paid $8.4 billion for the Finnish gaming studio Supercell, one of the largest tech deals on the continent ever. The company's CEO, Ilkka Paananen, told me: 'We sort of neutral countries in Nordics and companies coming from them could not build these truly global businesses that can have customers in every single country in the world . . . But it's becoming more challenging as the world is getting more divided. Of course, from an economic perspective it is a bit worrying.'

After decades of record growth, Chinese tech companies are now reeling from a dramatically altered operating environment both at home and abroad. Regulatory pressure is making it harder for them to acquire new assets and hold on to existing ones in the West: a place which had for decades been the sole focus of their efforts to go international. Tencent had to divest entirely from its 5 per cent stake in Tesla just as the EV manufacturer became the largest car company in the world.

The Chinese government has also been cracking down on the power of big tech. In December 2023, it introduced new restrictions on gaming companies, aimed at curbing the amount of time people spend playing video games. Tencent, once an almost trillion-dollar company, has seen its market value slashed by a third. Amid this tumult, Wallerstein, the most powerful American at the Chinese tech giant for most of the company's history, transitioned out of the CXO role in January 2024 and is now an advisor to the company on topics like climate and health.

I asked Wallerstein, the rock guitarist who bleeds Tencent, the native Californian whose personal story is woven into the founding myth of one of China's biggest tech giants, if he is at all conflicted about tensions between the country that he is from and the country that made him who he is; what he thinks about the negative halo around China tech; the current impasse; and where things might go from here.

'I don't think there's real architecting of this moment going on. I think there's some guidance, there's some things said at a high-level, announcements made which then can be reflected in policies, but I don't think there's a very clear vision of where people want things to go.

'It's not really clear how Chinese companies would be seen positively in the US for doing good things from a US perspective. To me, I think importing a lot of American IP, movies, sports, media, music, Snoop Dogg kind of stuff to China should get you a gold star. I don't know if there's an authority that gives gold stars, but I just didn't really on the ground ever get a pat on the back or a handshake, say, "Hey, at least you guys, when you import our movies or our music or our games or our intellectual property and pay us a fair rate for it, pay us money, that's good, that's positive."

'And I think governments around the world need to figure out what they believe is positive behaviour in addition to the negative behaviour. So you could say, *that's* bad, we don't like *that*. We think you've been stealing IP, that's bad. Don't do any more of that. But then when you do *that*, that's good, do more of *that*. And I encourage across borders, positive behaviour. I think that's entirely okay to say we like this behaviour, we want people to import more foreign stuff, our stuff or whatever. I think we need to get back to identifying where the carrots and the sticks are. And where are the carrots? Because I think it's just been nothing but sticks.

'I don't think people are really clear about where they're trying to go, where they are headed to. The best friends of America are all nations that America fought wars with. So the moment the bosses say, "Okay, we've agreed to something, we've negotiated, it's done," it's amazing how people then just stop and get along.'

Brains

In 2015, four researchers at the Microsoft Research Asia (MSRA) lab in Beijing published a seminal paper which marked one the most significant advances in machine learning. 'Deep Residual Learning for Image Recognition', or simply ResNet, outlined how neural networks can be layered to vastly improve the performance of AI systems.

Artificial intelligence is a general term that covers a range of approaches used to simulate human-like cognition in machines. Neural networks are one popular approach. Neural networks are a bit like artificial versions of the biological neural networks in our brains: a complex web of interconnected nodes, or neurons, working in concert to process information.

Researchers had long hypothesized that these neural networks can be layered to increase the computational power of AI systems. But implementing this in practice proved challenging. As the layers of neural networks got deeper, the quality of the signal passing through them got weaker. It's a bit like trying to pass a message through a long game of telephone. The more people the message passes through the more distorted it gets. In deep learning this is called the 'vanishing gradient' problem.

The MSRA team in Beijing solved this problem by introducing a technique called 'skip connections'. They worked out a way to train neural networks hundreds of layers deep without any apparent loss in signal. This paved the way for more powerful AI systems which enabled applications like facial recognition and autonomous driving. ResNet is also one of the core enhancements that separates AlphaGo, DeepMind's program which can play Go at a superhuman level, and AlphaZero, its more advanced game-playing bot with more generalized capabilities, which can play multiple games like Go, chess, and shogi – three of the world's most strategic board games – at a superhuman level.

The ResNet paper has in less than a decade racked up almost a quarter of a million citations on Google Scholar. It is the most cited artificial intelligence paper ever. It is also the most cited computer science paper ever. It is in fact the most cited scientific paper *in any academic field* published in the twenty-first century.

The ResNet paper was authored by Kaiming He, Xiangyu Zhang, Shaoqing Ren and Jian Sun. All four got their undergraduate, graduate, and doctoral degrees at Chinese universities. None of them had worked outside China prior to publishing their landmark paper. They have all won awards at top global AI conferences and computer vision contests.

In 2015, they won ImageNet, a major competition in artificial intelligence at Stanford which invites researchers to submit AI algorithms to identify objects in images. The MRSA submission was the first time that an AI system surpassed human-level performance at image-recognition on their dataset. ResNet was not a one-off. The last three editions of the ImageNet challenge were all won by teams that came from China.

'China used to produce a lot of poor-quality research in artificial intelligence,' says Geoffrey Hinton, acclaimed British computer scientist and professor at the University of Toronto. 'Then it started to produce a lot of poor-quality research and some good-quality research. Now it produces a lot of good-quality research and some absolutely world-class research.'

Hinton is the author of 'ImageNet Classification with Deep Convolutional Neural Networks', or AlexNet, which was the most cited paper in artificial intelligence before ResNet came along. It is still the second most cited scientific paper in the field. Hinton has also won the ImageNet challenge along with just about every major accolade in computer science, including the Turing Award, known as the Nobel Prize in computing, and, just for good measure, the actual Nobel Prize, this time in physics. He told me that China's prolific research output makes it the one country that is best positioned to take the lead in artificial intelligence. He opined that China's growing scientific prowess may very well make it the next global power whose scale could 'rival that of the British Empire'.

In 1990, China was responsible for only 1.2 per cent of the world's scientific publications. It has now surpassed the US as the world's leading source of scientific papers. Almost a quarter of the world's publication output in the sciences comes from China. It is not just volume. China also produces most of the world's highest quality research. In 2022, it surpassed the US in the *Nature* Index, a ranking of countries which contribute the highest number of papers to the world's most prestigious journals. It has also replaced the US at the top of a ranking released by Clarivate, a science analytics company, which tracks publication impact based on citations. A recent cover story in *The Economist* on the rise of China as a scientific superpower declared: 'The old science world order, dominated by America, Europe and Japan is coming to an end.'

2

Microsoft Research Asia, the lab that produced ResNet, was founded as Microsoft Research China in 1998 in decidedly modest circumstances at a time when less than 0.1 per cent of China's population was connected to the internet. The lab had only five employees who worked out of a nondescript office building in the Zhongguancun neighbourhood of Beijing which at the time was mostly villages and vegetable farms whose uneven streets were littered with ox-drawn carts.

Since then, the area has become the heart of Beijing's thriving tech community. Zhongguancun is now known as China's Silicon Valley and listed among the ten priciest places in the world to rent commercial real estate. Dotting its avenues are China's leading universities like Tsinghua and Peking University, and offices of China's biggest tech giants like Baidu, Sina and ByteDance. Over half of all of China's billion-dollar tech startups – more than a hundred companies – are based in this area.

ResNet is not the only breakthrough to come out of MSRA. The lab developed technologies for parsing and rendering digital ink, laying the foundation for pen computing. Much of the technology that goes into facial recognition systems used everywhere in the world also came from here. MSRA has been called 'the hottest research lab in the world' by the *MIT Technology Review* and 'the single most important institution in the birth and growth of the Chinese AI ecosystem' by the Paulson Institute. The Chinese media have called it 'The Whampoa Academy of the Chinese internet', a reference to the early twentieth-century military academy that churned out commanders for the Kuomintang and the Chinese Communist Party.

MSRA, which has now grown to over 200 scientists and 300 visiting researchers, is to China what Bell Labs or Xerox PARC were to the US in their heyday, widely considered something of a cradle for the Chinese tech elite. Its 7,000 alums include some of the most influential figures in the country's tech industry: Wang Jian, the Chief Technology Officer of Alibaba; Ying Ma, Chief Scientist and Head of AI at TikTok's parent company ByteDance; Yin Qi, founder and CEO of Megvii, a leading Chinese company working on facial recognition; Tang Xiao'ou, the billionaire co-founder and CEO of SenseTime, another facial recognition company; and Li Shipeng, a founding member of MSRA who now heads research at iFlyTek, one of China's most prominent AI companies.

Microsoft was one of the first major multinationals to start a major research lab in China and in doing so created a template that many would follow. Over 1,300 foreign companies have opened advanced research labs in China to tap into the country's growing pool of scientific talent, some explicitly modelled on MSRA and staffed with former MSRA researchers.

3

Ya Qin Zhang was one of the five founding members of MSRA. He is now a celebrated figure in the Chinese tech industry and carried the torch when the Olympics came to Beijing in 2008. But his story has improbable beginnings.

Zhang lost his father to the Cultural Revolution and was raised by his mother and grandmother. When he was a child, his mother thought her son wouldn't even make it to college. Chairman Mao regarded the educated classes as counter-revolutionary and shuttered the entire education system for a decade, from 1966 to 1976, to purge elitist and bourgeois tendencies from society. Mao died in 1976 and soon after schools and universities slowly flickered back to life. In 1978, Zhang took the first national level college admissions test offered in China in twelve years.

He passed the exam and made national headlines. Zhang was only 12 years old, the youngest person to be admitted to college in China at the time. The child prodigy enrolled to study engineering a thousand miles from home, at the University of Science and Technology (USTC), sometimes called China's Caltech, a small leading science school, in Hefei, the capital of Anhui province in eastern China.

When he graduated, Zhang landed a scholarship to study in the US where, at the age of 23, he finished his PhD at George Washington University. He became a fellow of the Institute of Electrical and Electronics Engineers (IEEE) at the age of 31, the youngest person to win this honour in the organization's hundred-year history. And the following year, in 1998, when he won the 'Outstanding Engineer of the Year Award', President Clinton sent him a letter praising him as 'an inspiration for others'.

Zhang spent sixteen years at Microsoft, including a stint as the Director of MSRA, and then the Chairman of Microsoft China, the company's highest ranking official in the country. In 2014, just as China's own home-grown tech companies were taking off, he left to become the President of Baidu where he took the lead on building out new technologies and business lines. Zhang now teaches at Tsinghua, a short drive from his old MSRA office in Zhongguancun, where he is the Dean of the Institute for AI Research, or AIR, China's most elite academic department for training AI talent.

I asked Zhang why, when most academics are leaving to take up lucrative positions in China's booming tech industry, he made the switch the other way around. 'Money wasn't really an object for me,' he says. 'When I looked back at my career the time I enjoyed the most was Microsoft Research.' AIR is Zhang's attempt to recreate the magic of MSRA in a university context where he can train the next generation of Chinese AI talent, which can make fundamental breakthroughs like ResNet.

Like MSRA, Tsinghua is also the product of a US–China partnership, though one that was forged under peculiar circumstances. It traces its origins to the Boxer Rebellion, the early twentieth-century uprising against foreign influence in China. After the rebellion was quashed by foreign powers, including the US, President Theodore Roosevelt obtained an approval from Congress to use some of the indemnities extracted from the defeated Qing dynasty to fund a scholarship programme to bring a hundred Chinese students a year to study in the US. Tsinghua College was set up in Beijing in 1911 with these funds. The US Bureau of Education and the YMCA recruited the first sixteen American teachers that made up the university's faculty. In their junior year, Tsinghua students transferred directly to American colleges to finish their studies in the US.

Born out of national humiliation, Tsinghua is now a major source of national pride. Widely regarded as the best university in the country, it is known particularly for its powerful alumni who play an outsized role in national politics. Graduates include the sixth and seventh General Secretaries of the Communist Party of China (CCP), Hu Jintao and Xi Jinping, often referred to as the country's 'paramount leaders'. The two Tsinghua alums have consecutively run China for an uninterrupted two decades. Given their alma mater's history, steeped as it is in the country's colonial past, and which traces its origins to one of the most humiliating episodes in recent Chinese memory, it is perhaps unsurprising that it is precisely these two figures who are most credited with leading modern China down a more assertive and nationalistic path.

Chinese universities have in recent years seen a surge in their reputation. According to the Leiden Ranking which ranks universities according to volume of research output, six out of ten of the world's top universities are in China. In the *Nature* Index, it is seven out of ten. Simon Marginson, Professor of Higher Education at Oxford, tells me that when judged solely on the strengths of their science, technology, engineering and mathematics departments, Chinese universities are now ahead of their American counterparts. 'Tsinghua is overwhelmingly stronger than anyone else,' he says. 'Number two is Zhejiang. MIT is about seventh or eighth, it's the only American one in the top twelve.'

Tsinghua's academic standing has been buoyed in no small part by returnee scientists like Ya Qin Zhang who are bringing what they've learnt at top research institutes in the US and China back to the classroom. 'If you look at the faculty, it's top notch,' Zhang tells me. Other prominent Tsinghua faculty members include mathematician Shing-Tung Yao, winner of the Fields Medal, previously a named professor at Harvard, and Andrew Yao, a computer scientist who has won the Turing Award. Yao left a named professorship at Princeton and renounced his US citizenship to return to China and take up a place at Tsinghua and the Chinese Academy of Sciences (CAS).

Zhang says that his vision to create another MSRA is working out better than expected. 'Tsinghua students, even undergraduate students, they're pretty good, it's so natural for them to be a part of a research team,' he says. 'The calibre of students is just as impressive as at MIT, Caltech and Columbia – it's amazing. I thought it would take longer, but I'm glad to see that it's happening.'

4

MSRA and Tsinghua are both products of the US–China partnership; just two prominent examples of what has historically been a much broader collaboration in science and tech. According to a paper in the *Journal of Quantitative Science Studies* published by the MIT Press, China and the US are each other's largest partners in high-impact research publications.

This collaboration is now under strain. Microsoft has been under pressure to pull the plug on MSRA; it has had to publicly reassure regulators at home that it has put in place guardrails to restrict researchers from politically sensitive work. Google has already closed its AI research lab in China. Many MSRA alums, like Li Shipeng, one of the institute's founders, are under direct sanctions by the US government. One of the ResNet paper's authors, Xiangyu Zhang, now works at Megvii, a prominent Chinese AI company, which has also been blacklisted by the US government.

It is still unclear whether these measures are succeeding in their attempt to curtail China's development of advanced research capabilities. What is clear though is that they are speeding up what was already an increasing trend towards an indigenization of research and development in China.

Jie Tang is one of Tsinghua's star graduates. He gained his doctorate there in 2006 and is now a full professor in the university's computer science department. His research focuses on Large Language Models (LLMs), a particularly hot subfield in machine learning. Jie has developed some of the most advanced language models to come out of China which power ChatGLM, the Chinese variant of ChatGPT.

In 2019, Jie launched Zhipu, an AI startup spun out from Tsinghua. He tells me that the move was motivated less by commercial reasons and more because universities have become less suited to big research. 'Even in the United States, even Stanford and MIT, they cannot do the same kind of AI research as OpenAI,' he says. 'Large Language Models need quite a lot of resources, that's the reason we built the company.'

Zhipu is the best funded OpenAI rival to come out of China. In the summer of 2024, it was valued at $3 billion. It is backed by some of the biggest names in China tech like Alibaba, Tencent and Xiaomi and raised a massive $400 million from the venture arm of Saudi Arabia's Aramco.

Zhipu has the same product line as OpenAI: a chatbot, a text to image generator and a text to video generator. Most people I spoke to seemed to think that Zhipu trailed OpenAI by about twelve months – whatever capability OpenAI had, Zhipu would also have it within a year. I asked Jie if his competition with OpenAI goes beyond just chatbots and image generators. 'Short term we'll just catch up with OpenAI; what OpenAI does, we'll do,' he says. 'For the long term of course, we will target artificial general intelligence.'

Zhipu's founders were trained in China. A significant proportion of its capital is denominated in yuan and not dollars. This is a departure from the norm among the previous generation of Chinese tech companies which tried to recruit foreign-trained talent and showed a strong preference for foreign capital. It's a telling sign that the country's tech industry, which hitherto looked abroad for resources and validation, is now becoming more indigenized. US investors had already been scaling back their investments in China. Now Chinese companies are also wary of taking US funds.

'I think it's very difficult geopolitically,' says Rui Ma, the COO of US-based market research firm Alphawatch, and a close observer of tech startups in China. 'People don't want USD. People want to keep their cap tables [ownership structure of a company] to be no US investors. They want to go public in the Chinese capital markets. I was talking to a company, they have enterprises as clients, and those enterprises don't want to use service providers that have US investors. It's kind of an odd thing but that's just the reality. Maybe it'll change because it's not an official rule, it's just people are like "Oh, I think that could be risky." So it is happening on both sides.'

Some of the brightest minds in China, who once used to make a beeline for universities in the US, are now rethinking their options. According to the US State Department, the enrolment of Chinese students at US schools has dropped by 20 per cent since 2019. A lot of that has to do with the environment of suspicion that now surrounds students of Chinese origin on American campuses. But there are also other factors at play.

'To me personally this is because people have a sense of belonging, a sense of civilizational pride,' says a well-known Chinese tech executive, himself a graduate of top programmes in the US and China.

Jie Tang lists on his website dozens of students he has supervised who have made their way to PhDs in some of the most competitive computer science departments in the world. Harvard, Stanford, MIT, Princeton, CMU, Berkeley, there's hardly any top tier programme that's not listed. I ask him why, when so many of his students have left, he decided to stay.

'Some of my friends tried to persuade me to come to the US

and work in a big company and even join some kind of university to work as professor,' he said. 'But finally, I mean, it's a choice, it's a decision. I thought I could do something big in China and also in that way maybe I can help China better. So I decided to stay.

'From the beginning, when I decided to stay at Tsinghua I was sure that Tsinghua students are actually the same level as Caltech, MIT and Stanford. But the difference is their resources and opportunities are not at the same level. So once you give them enough opportunities to visit different companies in the world and different universities, and if you give them the same resources, I think they can grow up very quickly. So that's what we did in the past ten years.'

Deployment

Basic research in China is in a much better place than it has ever been. But it would be prudent not to overstate its significance relative to other countries. The overall volume and quality of research output has risen sharply. But few of these papers cross the threshold where they can be considered truly groundbreaking discoveries and innovations. ResNet is still the exception and not the norm of the type of research coming out of China.

By some measures China is generations behind Western nations. In 2023, the US produced as many Nobel laureates in the sciences in a year as China has in its entire history. Only five people from mainland China have ever won the Nobel Prize in the sciences – four of whom got it for work done at universities in the US. This suggests that even if China's overall share of highly cited research in quality publications is on the rise, the US continues to produce a disproportionately higher share of influential, groundbreaking works relative to its total output.

There are two interpretations of China's poor showing in winning

Nobels. One, offered by Simon Marginson at Oxford, is that Nobels are a trailing indicator of scientific achievement. 'The Nobels cycle is decades long and Chinese science in its contemporary flowering is twenty years old,' he says. 'Nobels are going to people who did their best work twenty or thirty years ago – I'll be surprised if China doesn't start to do better.'

The other interpretation is that the culture of US research is inherently more suitable to producing groundbreaking discoveries. The US nurtures fundamental scientific enquiry while China's research ecosystem remains more oriented towards applied research. It will take decades to change those deeply ingrained institutional tendencies.

'Truly breakthrough technological innovation, at least in my own estimate, for two generations won't happen in China,' says a well-known figure in China's tech industry, who has held high-ranking positions in US and Chinese tech giants. 'The odds are very low. Because the United States, in my view, is sort of unique because of the immigration culture and university environment.'

Most Chinese tech figures readily concede that China has a lot of catching up to do in producing fundamental breakthroughs on a consistent basis. And they're fine with that. In their view, China's near term innovation advantage is not in developing new technologies but in adopting them faster than anyone else. They see the arrival of new technologies playing out in two phases. The first is invention. Which is about knowledge and creativity. Here the US is ahead and China is behind. The second is deployment. Which is about speed, efficiency and discipline. Here China has an edge.

Kaifu Lee, the founder of MSRA, who has been the highest-ranking official for both Microsoft and Google in China, and is now the Managing Director of Sinovation, a leading venture fund, told me: 'So the transformation that China went through is: from copycat to micro innovation to actual real innovation. So basically three phases. Today I think the state of the world is that brilliant new ideas still largely come from the United States, but the Chinese have a stronger work ethic, engineering discipline, execution capability, tenaciousness and willingness to do whatever it takes, even if it's boring, to build a successful final product. So it's a combination of tenacity, hard work and willingness to put up with boring stuff. And I think that is an incredible combination for building a successful product.

'And while the most amazing discoveries are American, from deep learning to convolution neural networks to more recently transformers, large language models – all American – I would argue their engineering execution will be done possibly better in China in terms of engineering excellence and in terms of product market fit, in terms of being humble and hardworking and continuing to hone the product, always feeling like I'm going to eat my own lunch because if I don't someone else is, as opposed to becoming complacent.'

If the central idea animating China tech in the past was to copy what already works in the US and bring it to a large captive market at home, the new playbook is to swiftly adopt new technologies, even if they're developed elsewhere, and deploy them at scale before they have been put in play anywhere else in the world.

Some examples are well-known: 5G, mobile payments, solar power, electric vehicles, industrial robots, lithium-ion batteries. None of these were invented in China. But for each of them the scale of adoption on the mainland now exceeds the rest of the world combined.

Other examples are still emerging, like autonomous vehicles. The US is still ahead of China when it comes to the sophistication of its autonomous driving technology. But China has a lot more self-driving cars on the roads. Beijing and Shanghai have hundreds of driverless taxis roaming their streets, a sight that is still rare in Western cities. Baidu operates the largest fleet of robotaxis in the world and its fully autonomous ride hailing service, Apollo, has a network of thousands of driverless taxis that extends across a dozen cities in China.

In Xiong'an, a city built from scratch 100 kilometres south of Beijing, the entire transport infrastructure is designed from the ground up for autonomous mobility. It has in its public transport system a fleet of autonomous buses in which rides can be booked in advance free of charge with a simple tap on the phone. The city, built on a site that as recently as 2017 was mostly swamps and agricultural land, is now seen as something of a sandbox for testing out new technologies before they are rolled out into other parts of the country.

Xiong'an is yet another example of China's seemingly endless capacity to summon bustling metropolises out of thin air. It is seen as a successor to places like Shenzhen and Pudong which didn't so much evolve but were willed into existence by fiat from the very top of the country's political hierarchy. Shenzhen is closely associated with Deng Xiaoping, who designated it China's first Special Economic Zone, a laboratory for testing market-based principles, a still controversial idea in 1979. What seemed like a minor experiment in economic planning back then has in hindsight turned out to be one of the most monumental decisions taken in modern China.

Deng's decision to enclose Shenzhen in an economic parenthesis thereby abjuring it from the socialist orthodoxy of the rest of the country marked the pivot, the first tentative step in opening up the economy and establishing the basic doctrine that now underpins the wealth of the country. And it was Jiang Zemin whose direct involvement changed Pudong from a once undeveloped district to a leading global financial centre. The city's skyline, marked by the distinctive spheres of the Oriental Pearl Tower and the bottle-opener look of the Shanghai World Financial Centre, has become the most frequently deployed visual metaphor for rising China.

Xiong'an is another Chinese president's signature attempt to instigate an urban miracle. Often dubbed 'the city of the future', it is seen as something of a monument erected by Xi Jinping to mark his time in office. Xiong'an is Beijing's spillover city, designed to pull functions away from the country's crowded capital, and in less than seven years has accumulated more than a million residents, around the same as Dallas or Amsterdam.

'I really think China can play a strong role in the second phase of any new technology revolution,' Kaifu continued. 'The Americans will generally invent most technologies. Let's say 70 per cent. The rest might be divided between Europe, China, India, other places. But what's important is that once the important technology is written up and available on the internet or in papers, Chinese companies are more likely to run away with it with this engineering excellence.'

Wrapped up in this conviction that even if China cannot out-invent it can still out-execute is a thinly disguised belief shared widely in China's technology circles that years of rapid growth with no meaningful outside competition has made US tech companies soft and complacent. In his book, *AI Superpowers*, Kaifu, who grew up in Tennessee, writes that Chinese entrepreneurs are like 'gladiators' who have absorbed 'the lessons learned in the Coliseum' to 'kill or be killed'. These spartan origins might not lend themselves to 'lofty thinking' but have bred a maniacal work ethic which makes 'the valley's companies look lethargic and its engineers lazy'.

I asked a top Chinese tech executive, who has worked in senior roles in big tech companies in US and China, how his experience of working in American tech companies compares with what he sees on the mainland. 'It's an interesting blend,' he said. 'On the one hand it's very Chinesey: corruption, kickbacks, you deal with a lot of people issues. On the other hand, you have a tech stack and engineering staff pretty much like Google. Some portion of the tech stack is honest to god better than Microsoft.'

It's not just about the scale of deployment. It's also about its speed. Just about everyone I spoke to thought that China doesn't just work harder but also runs faster. Eight out of ten startups that have been the fastest to reach a billion valuation are from China, all of which got there in less than eighteen months. China went from having no high-speed rail to having twice as many tracks as the rest of the world combined in less than a decade. Temu, the low-cost alternative to Amazon, was launched in September 2022. By 2023, it was the most downloaded free app in the US and the second most visited e-commerce site in the world. Just a year into its launch its sales topped $5 billion.

'So the difference is that Chinese companies, even the big companies, are very fast – agility in implementation,' says Ya Qin Zhang. 'And also just very fast in terms of getting into new areas. When I joined Baidu their main business was search. After I joined we got into cloud computing, we got into autonomous driving, we got into all of AI, digital assistance, we got into silicon, new chips. Baidu develops its own chips. Half of Baidu's infrastructure like cloud or AI is using its own chips. Just like Google. We got into quite a few new businesses and we made that decision very quickly. Obviously there's a lot of work, strategic analysis, but it's very, very fast.'

I pressed Zhang, who has been both the Chairman of Microsoft in China and the President of Baidu, on how, in his experience, this agility plays out in practice. 'Let me give you an example,' he says. 'When I was at Microsoft we had big meetings. In Beijing we have two big buildings. I wanted to do a simple thing. Every time people come to a meeting, they have a laptop, they have to connect with the projector, it takes five minutes, ten minutes, a lot of time, just a long delay. So I said, "Why don't we just make a wireless display?" That should be straightforward. Then of course we came back with a proposal. We had to form a team. And just for months it didn't get anywhere. It's not a big thing. But they just couldn't do it.

'When I joined Baidu, I asked the same thing from the IT team. And within two weeks we had all 500 conference rooms with fully wireless projectors. Not only from your PC but also from your phone. They made it so simple. You just have a code. It's that easy. They didn't come back to me with a dedicated team or a project plan. They just did it. I think that is true for a lot of Chinese companies. They see something and a bunch of people just get it done. It's a culture of agility. People get to implementation very efficiently.'

2

Companies trying to roll out new technologies wouldn't get very far if the public isn't receptive to that change. In early 2024, just as Baidu announced its sixth generation robotaxi, a Waymo driverless car was attacked by a mob and set on fire in San Francisco's Chinatown neighbourhood. There have also been incidents of pedestrians punching Waymo's vehicles and slashing their tyres. In 2020, there were over a hundred incidents of arson and vandalism against 5G towers and other wireless infrastructure across the UK because of a conspiracy theory that radio waves sent by 5G technology make people's bodies more susceptible to the coronavirus. When electric scooters first made an appearance in San Francisco it didn't take long for anti-scooter types to throw them off parking garages, set them on fire and dump them in the lake.

That kind of extreme techno scepticism and a general resistance to newness and change aren't really a feature of today's China. Is that because people are just not at liberty to express dissent? Maybe. But it's also in no small part because the Chinese mindset has been moulded by an environment where rapid change is just the norm. Imagine someone born in Shenzhen in 1980. They are now comfortably in their midlife but by no means old; a full half of their lifespan still lies ahead of them. And yet within half a lifetime they've witnessed their surroundings change at a pace and scale that few other people would have experienced in human history.

In 1980, Shenzhen was a small fishing village with 30,000 people. An average resident made less than $15 a month. If the place was known for anything at all back then it was for being the exit route for illegal immigrants leaving China to find better opportunities across the river in Hong Kong, then still a British colony. Over half a million people left communist China via Shenzhen in a manner strikingly reminiscent of migrant crises along the coasts of Europe today. Some left on makeshift rafts, others braved bad weather and sharks to swim across the Shenzhen Bay, which is three miles at its narrowest point, often only to be arrested on arrival by the Hong Kong border patrol.

The flow of migrants now runs the other way. On 30 June 2024, the Chinese government opened the Shenzhen–Zhongshan Link to connect the two sides of the Pearl River Delta, slashing travel time between Hong Kong and the mainland by half. The 24-kilometre, eight-lane highway, a megastructure that holds ten world records, spans two suspension bridges, two artificial islands, one of which is shaped like a diamond, and an undersea tunnel. On its opening day over 125,000 cars packed the highway from Hong Kong to Shenzhen, causing long traffic jams. The journey from Hong Kong that is supposed to take just ninety minutes on this day took six hours. That weekend alone over a million Hong Kong residents made their way to the mainland. Forty years ago, mainlanders would risk their lives to escape to Hong Kong. Today the overall cross-border flow between Hong Kong and Shenzhen tilts towards a net inflow into the mainland.

They're coming mostly to access Shenzhen's booming economy. The former fishing village is now a busting metropolis of 17 million people, the third largest city in China after Shanghai and Beijing.

It has the second highest number of skyscrapers of any city in the world, the fifth highest concentration of billionaires, and the seventh highest concentration of Fortune 500 companies. The city is of course best known for its tech sector and is home to Huawei, Tencent, DJI, and BYD. Shenzhen, which practically didn't exist forty-five years ago, has more billion-dollar tech startups than all of Germany, a country that has been around in its modern form for over a hundred and fifty years. Residents in this 'city without a history' are on average the youngest and richest in all of China.

Zak Dychtwald, an expert on young China, says that it is this rapid economic change that has shaped the Chinese public's unique attitude towards adoption. He uses what he calls the Lived Change Index to show how the velocity of change in China is unlike that in any other country. The index uses per capita GDP to track how much economic change a population has experienced over their lifetime. Since 1990, Americans have seen per capita GDP grow by roughly 2.7x. In contrast, someone born in China in 1990 has experienced per capita GDP grow by 32x, and Shenzhen, 322x.

In 1980, when Deng Xiaoping first started opening up the economy, China's share of the global GDP was 2 per cent. In 2024, it is 18 per cent. 'To have lived in China since 1990, broadly speaking, is to have lived in a country that is moving faster and changing more quickly than any other place on earth,' writes Dychtwald. 'You might ask yourself how living through that sort of change would shape your expectations for progress and your sense of what government, technology and commerce can do.'

So if the Chinese public gets more technology, it's in large part because it wants more technology. There's a strong element of Keynes's Law at play here, where more demand creates more supply. This has given the Chinese consumer access to products and services which are in many cases better than what's on offer elsewhere and in some instances not available anywhere else at all.

'I think in consumer experience, China is better almost in every aspect,' says Rui Ma, founder of a market research firm based in California. 'It's super ultra-competitive and people are just very demanding. When people visit the US from China, they generally say the service is really bad and things are very slow.'

Starting ten years ago, most major online shops offered same-day delivery in major cities in China, and it's free, it's included. That's still not at all common in the rest of the world. Facial recognition is built into all sorts of services to reduce friction and make things go faster; at airports instead of showing their boarding pass, even the mobile version, passengers can just scan their face and board a plane.

There is also a much tighter integration between online and offline in China. For at least five years, Chinese shoppers have been able scan QR codes on products like meat and produce in grocery shops and see the entire supply chain from farm to shelf right there on their phones. This blurring of boundaries between the physical and virtual, happening across industries in China, has been slow to catch on in other markets. And it's not because others don't have access to the same technology. The difference largely comes down to how people in other countries are just not as receptive to new technologies.

3

The relentless culture of Chinese tech companies and the fiercely adoptive nature of the Chinese consumer go some way to explaining why China is an outlier at embracing new technologies. But this discussion would be incomplete without the Chinese government. Here the picture is more complex.

The Chinese government is in some ways a major benefactor of domestic tech companies. It provides them with substantial financial assistance in the form of subsidies, tax incentives and direct investments. For instance, according to the CSIS, Beijing poured $125 billion into the EV sector alone between 2009 and 2021. Chinese companies also benefit from state support in indirect ways. 'China has a tremendous capacity for centralized achievement of objectives,' notes Simon Marginson at Oxford. In other words, its regime possesses a quality rare among governments: competence. 'I would say the government in China is very, very, very good at building infrastructure,' says Rui Ma. Chinese companies have the advantage of building their services on top of this strong and constantly evolving base layer, which gets things moving quickly.

And it is this competence which in no small part upholds the government's legitimacy. The central paradox of contemporary China is that the government maintains a highly controlled political structure while also enjoying an unusually high degree of public confidence. According to a recent study of Chinese public opinion spanning fifteen years, 95 per cent of Chinese citizens expressed their satisfaction with the government. Those look like North Korea numbers, except the survey was conducted by the Ash Center at Harvard's Kennedy School of Government. 'We tend to forget that for many in China, and in their lived experience of the past four decades, each day was better than the next,' Tony Saich, the Director of the Ash Center who led that study, has said.

Even the most uncharitable polls usually put public support for the Chinese government between 50 to 70 per cent, at par with the approval ratings of the most popular governments in Western democracies, like President Obama's when he left office. In an article titled 'What the West Gets Wrong About China', Rana Mitter, a Professor at the Kennedy School, and Elsbeth Johnson, a Senior Lecturer at the Sloan School of Management, write: 'Many Chinese believe that the country's recent economic achievements – large-scale poverty reduction, huge infrastructure investment and development as a world-class tech innovator – have come about because of, not despite, China's authoritarian form of government.'

A well-known Beijing-based tech executive, who left China in the 1980s to study in the US and spent decades in senior roles at two major American tech companies, told me that he never thought he would ever go back to China, 'but things change to be honest,' he said. 'I'm a non-political person, but I thought that China as a nation, the economy, all my relatives, my parents, they're now happy. Their living condition honestly is very good. And I also have a slightly different view from my initially hostile view towards the Chinese government. These people, they do a pretty decent job. I shouldn't hold a grudge against them because I had a pretty poor childhood growing up, because if you look at people's daily life compared to what I was used to before, it's changed, and I think they actually deserve a lot of credit for doing this.'

But the Chinese government's expansive presence also in some ways hinders technological change. Even if companies like BYD, Tencent, and Huawei are corporate China's most recognizable faces abroad, at home it's still state-owned firms that hold sway over the economy. State-owned enterprises account for 60 per cent of the market cap of listed firms in China. All ten of the country's largest corporations and three out of every four Chinese companies in the Fortune 500 are state-owned enterprises. This crowds out private initiative from large swathes of the economy, making entire sectors effectively off-limits for creative destruction by new startups.

The state's presence is also felt in sectors that are not under its direct control. For instance, the Chinese government tightly controls the Initial Public Offering (IPO) process. In the US, going public is an application-based process; in China it is approval-based. Unprofitable companies are simply not allowed to debut on the stock market, a mindset that would seem old fashioned in the US. Airbnb and Amazon were considered highly successful public companies even if for over a decade they remained highly unprofitable businesses. In China those sorts of business models simply wouldn't be allowed to enter the public markets.

'China has its own speed and rhythm, so you need to really understand it,' a prominent Chinese venture investor told me. 'You cannot just say, "Okay, let's try to create a bunch of space tech companies!" You cannot do that. Policy is really important. So that's why, it sounds weird, but we have to watch CCTV news every day and we have to judge. And we have to study the résumé of each bureaucrat and figure out which sector might benefit from those new promotions and designation of new officers.'

Chinese companies have learnt to their detriment that what can be rolled out quickly can also be rolled back quickly. Examples abound. In November 2020, Beijing pulled the plug on Ant Group's highly anticipated IPO which was expected to value the company at over $300 billion, the largest stock market debut in the world at the time. In subsequent years, Chinese tech companies have faced unprecedented scrutiny. Alibaba was slapped with a record $2.8 billion fine, and Didi, $1.2 billion. The crackdown wiped out a trillion dollars in value – the size of the entire Dutch economy – from Chinese tech firms.

It is now abundantly clear to corporate China that when the gears of the Chinese bureaucracy move in the other direction the fate of entire industries can turn on a dime. In 2021, cryptocurrency mining and trading were made illegal. Then for-profit tutoring was banned, sinking the entire online education industry practically overnight. Soon after, new restrictions were introduced on videogames, which state media outlets called 'spiritual opium', limiting gaming time for players under eighteen to just three hours a week, and not just any three hours: 8 p.m. to 9 p.m. on Fridays, Saturdays, Sundays and public holidays. The announcement sparked an $80 billion market meltdown in the gaming sector.

What were the underlying motivations for what was called a 'summer blizzard' or 'crackdown of everything'? The most benign interpretation is that at least some of the measures were justified regulatory oversight. The crackdown on private education was applauded by many as a necessary corrective to ease the pressure of the rat race – known in China as 'involution' which means 'one does not grow or progress but merely spins in place' – that was depriving the young of their childhoods. Zhang Taisu, a professor at Yale Law School, has called it a 'natural, long overdue course-correction'. Some have interpreted the expansive reach of the crackdown in political terms: the country's old political power putting its new money power in place. And yet there are others who divine broader ideological and grand historical currents at play, i.e. the return of Red China.

Perhaps no single phrase captures China's de facto ideology in the past twenty-five years better than Deng Xiaoping's pronouncement 'Let some people get rich first'. That Gilded Age, some say, may have reached its climax. A recent paper by Thomas Piketty, Li Yang, and Gabriel Zucman finds that China has now joined the club of rich capitalist countries in measures of wealth distribution. 'China's inequality levels used to be close to Nordic countries and are now approaching US levels,' it concludes. In influential branches of the CCP this is seen as a signal that the necessity of Deng Xiaoping's concession to private enterprise has run its course and it's time now to switch gears.

Soon after the crackdown began, in October 2021, President Xi Jinping wrote an essay for *Qiushi*, the CCP's flagship theoretical journal, which invoked the idea of 'Common Prosperity', an ideologically loaded term that dates to the time of Chairman Mao. The phrase means different things to different people but captures a tendency, an entire way of thinking, like 'Make America Great Again' or 'Take Back Control', that in its current incarnation is widely understood to encapsulate the party's intent to curb what Xi Jinping has often called 'barbaric growth' and 'disorderly expansion of capital'.

But the Chinese government's domestic preference to curb the excesses of capitalism might not align so neatly with its foreign policy objective of rolling back US influence in Asia. While the current government has drawn a sharp distinction between 'Chinese modernization' and 'Western modernization' to guide economic thinking at home, it has also pushed for a 'whole nation' approach – another echo of mid-twentieth-century communist thought – to mount an effective technological challenge to what Xi Jinping has called US-led efforts at 'containment, encirclement and suppression'. It is hard to see how decapitating the country's tech sector, the component of the 'whole nation' that has thus far delivered said technological progress, would serve that objective.

That realization is not lost on the Chinese government. The crackdown that started in 2021 is now seen to have passed with the big tech companies largely falling in line. Some companies ceded small ownership stakes to state entities, known as golden shares, which gives the government formal say in business decisions. It remains to be seen whether the tech sector's uneasy dance with the state authorities will be enough to ensure that in the coming years the policy pendulum does not swing too far from 'let some people get rich first' to 'common prosperity'. The answer may lie in whether the companies can deliver the one thing the party could possibly want more than a domestic triumph of the Chinese system over Western-style capitalism: an unassailable lead for China in the global contest for new technologies.

Kaifu Lee has been a teacher, researcher, tech executive and venture investor. As our conversation drew to a close, I asked him, what comes next? 'Venture building,' he said. I understood what he meant a few weeks later when news came that he had launched 01.AI, a Beijing-based startup that competes directly with OpenAI. In the summer of 2024, Stanford certified the new company's large language model, Yi-Large, as the third most advanced in the world. 01.AI, called Ling-Yi Wan-Wu in Chinese, alludes to a passage from the Taoist text *Tao Te Ching* which means 'Zero-One, Everything'. It reached a $1 billion valuation in eight months.

'China is no longer a copycat, it is the student,' Kaifu told me. 'America is the teacher, China is the student when it comes to great technologies; but China is now a precocious student who is learning from the teacher but also maybe out-executing the teacher, right?'

CHAPTER TWO

Steeples of Excellence

'I've learned not to bet against Silicon Valley because I already did that once and I was wrong.'

Is it still the Valley?

China's rise in the ranks of technologically capable countries is no longer in dispute. There are disagreements about whether that's a good thing or a bad thing, but there is broad agreement on the fact that it's a thing. Ask ten people if they think China is a plausible competitor to the US in tech and nine will say yes and the tenth will say it's already left it behind; and, ironically, the ones prone to overstating the country's capabilities the most are typically the ones who like it the least. The natural corollary to that is: is the US in decline? Here the picture is uncertain.

Is the centre of gravity of where the most consequential developments in tech are taking place moving outside the US? Is it staying in the US but moving to more places outside the San Francisco Bay Area? Is the Bay Area's relevance going up? Is it going down? Is the gap between the US and the rest of the world widening? Is it narrowing? Unlike the question of China's rise, there's hardly anything approaching even close to a consensus on any of these questions. I came across equally formidable authorities who made an equally compelling case one way or the other in response to each of these queries. So here I'll try and make sense of all the conflicting positions I came across about what's going on out there and why.

Let's start with the naysayers. People have been writing obituaries of Silicon Valley for just about as long as it's been around. But it does seem like that chorus has been growing louder lately. 'Silicon Valley is over,' declared the *New York Times* in 2018. 'Peak Valley,' said *The Economist* soon after. And a particularly sharp critique, 'The end of the Silicon Valley myth', that appeared in *The Atlantic* in 2022, observed: 'The tech giants that have shaped our lives, online and off, over the course of the 21st century have at last hit a wall . . . Now, ruled by monopolies, marred by toxicity, and overly reliant on precarious labor, Silicon Valley looks like it's finally run hard up into its limits.'

It's not just the media, tech insiders have also been sounding the alarm. Michael Moritz, a former top partner at Sequoia, took to the pages of the *Financial Times* to air his misgivings about where things are headed. In 'Silicon Valley would be wise to follow China's lead', an opinion piece published in 2018, Moritz complained that while Chinese companies were pulling ahead with their ferocious work ethic, US tech workers were distracted by 'soul sapping discussions' about 'the inequity of life'. It was a rebuke to both the far-left politics that has made its way into tech companies and the coddling of the American tech worker who has become more about work-life balance than shipping the next killer app. Moritz lamented that these were all signs that society had become 'unhinged' before concluding that 'doing business in China is easier than doing business in California'.

This runs parallel to a wider strain of thinking that the Valley's intellectual environment has been hijacked by identity politics and the place has gone too woke. Fringe social issues that cater to narrow interest groups are dictating terms for mainstream politics at the expense of more utilitarian matters that concern everyone. Too much talk of social progress and not enough of the economic and technological kind. The zeitgeist has shifted too far in the direction of equity over productivity, social justice over social order, and cultural diversity over social coherence.

But then there is the left-leaning counter-narrative that argues the opposite. The problem isn't so much that the Valley has become too woke as that its thinking has become more market-oriented than socially driven. This has dampened the region's progressive streak – being two steps ahead of the rest of the country on all things social justice – which made the Valley what it is in the first place. Creativity and divergent thinking are just a by-product of these counter-cultural values. Take away the Valley's demons, they say, and its angels will flee as well. 'Social diversity and common good' are what distinguish true San Franciscans, David Talbot, the founder and former editor in chief of *Salon* has said. Adding that the current conflict 'pits San Francisco's bedrock progressive values – including a strong commitment to social diversity and the common good – against the defiantly individualistic, even solipsistic, world of digital capitalism'.

All of which is to say that the Valley has been under assault from both sides of the ideological divide. And while their claims about what ails the region – too much progressivism or too little? – might seem mutually contradictory, the red thread that runs through the two competing narratives is that the place just isn't what it used to be.

'I think the Valley's fundamentally changed, the culture has shifted,' says Victor Hwang. Victor spent almost a decade in the Valley as an entrepreneur before leaving for Kansas. He tells me that he left because the place is now less about the dreams and more about the commerce, a place where the missionaries have largely given way to the mercenaries.

'I kind of saw it when the Valley jumped the shark and it was when people that were moving there, they really weren't lovers of innovation itself. The game really shifted towards how do I make money? And how do I build my own personal brand? How do I extract value from the Valley as opposed to how do I add to the Valley? It changed because the people that started coming changed. And now you see so many people leaving because the magic has started to fade away.'

Victor, who has been a VP at the Kauffman Foundation, a nonprofit that supports entrepreneurship, best known for launching the Kauffman Fellows Program, a well-regarded training programme for young venture capitalists, is the author of *The Rainforest*, a book about what living in the Valley taught him about why some places are more creative than others. 'What I was capturing at the time, and I had a sense of it, I had a sense I was there at the peak of Florence, and I was trying to write it down in part too because I knew it was a historical moment that might be gone,' he says. 'And so I was trying to say: this is what it feels like to be in Florence at the height of the Renaissance.'

The book is a meditation on how some environments are like farms and other environments are like rainforests. Farms are controlled systems where inputs and outputs are predetermined. You plant these seeds, you get those crops. Schools, companies and governments are all these farm-like environments. In these settings there's a firm distinction between what you want, crops, and what you don't want, weeds, and the whole thing is optimized to give you a lot of one and none of the other.

Rainforests are a lot more chaotic. Nothing is predefined: inputs, outputs, or even processes. A diverse set of inputs come together in unusual combinations to churn out lots and lots of differentiated outputs, an environment primed for the evolution of new species. In rainforests there's no distinction between crops and weeds; what may look like an undesirable weed today could turn out to be the most valuable new crop tomorrow.

Victor thinks that the Valley, which used to be this massive entrepreneurial rainforest, has over time acquired more farm-like characteristics. The experimentation and divergent thinking that fed its creative instincts have long been replaced by a more commercially minded focus on producing high-value companies.

'I will say I think rainforests don't die overnight and so there may be a few more bounces in the Valley, but the long-term decline is clear and you can see that in people,' he says. The bright side is that even if the Valley is fading, the culture that made it what it is has found its way to more places. 'My hope is that becomes Silicon Valley's greatest legacy: that it taught the world how to remember to innovate. And before the Valley itself disappeared, it threw out that message to everybody else: here's how you do it.'

2

High-profile departures of major tech figures from the Bay Area have fuelled the perception that the Valley's best years might be behind it. Elon Musk left for Texas, Peter Thiel for Los Angeles, Alex Karp for Colorado, and Larry Ellison for Hawaii. Other destinations, from the decidedly very happening Austin to the not so plausible Raleigh-Durham, have been propped up as 'the next Silicon Valley'.

Keith Rabois, Managing Director at Khosla Ventures, is one of the prominent figures to exit the Valley. I asked him why and what this meant for the future of tech in the US.

'I think Silicon Valley is going through its Detroit correction, but I don't think technology is,' he said. 'Technology has just the brightest future today as it did five years ago, ten years ago, twenty years ago. The world still needs more tech. Tech solves problems and the world has lots of problems to solve. And the only magic wand to solve most problems in life is through tech. And so that's going to continue. But I think places and the people that solve problems through tech are going to be more widely distributed and less likely to be stuck in Silicon Valley in the Bay Area.'

Rabois has been very vocal about the rise of Miami as a rival tech hub. For him it was a question of the city just being a lot more liveable compared to its more established competitor out in the West. 'At the end of the day, it relates to being happy,' he said. 'Fundamentally it's very difficult to build a company when you're constantly being assaulted, your property is being stolen, and homeless people are accosting you. So the more distracted you are by government failures, the less likely you are to concentrate on your work. So I think Miami being a safe, vibrant, exciting city allows people to do their best work.'

It's a message which I heard often. It's not so much the decline of the US as it is the decline of the Valley, with the action moving to other centres in the US: Austin, Miami, LA, Boston. But the one alternative that comes up most often is New York City. The city has long been the second most important tech hub in the US, ranking behind only the Valley in all the metrics that matter: number of billion-dollar companies, valuations, exits. The city has produced its share of headliners, like Etsy and Warby Parker.

Chainalysis is among the more recent New York-based tech startups making waves. Launched in 2014 as a digital forensics firm for the blockchain era, the company builds software that can trace blockchain transactions, like bitcoin payments. The company has an interesting backstory, so it would be useful to take a detour through its origins.

The startup was born out of one of the biggest blow ups in the crypto space. Most people woke up to crypto exchanges and how dodgy their underlying business can be when FTX, one of the world's largest crypto exchanges, imploded in 2022, wiping out tens of billions of dollars in customer deposits and landing its founder Sam Bankman-Fried in jail. But before FTX there was Mt. Gox, the OG crypto disaster.

Mt. Gox was a crypto exchange that emerged in Tokyo in 2010

back when bitcoin was still just a rumour. On the first day of operations, 20 bitcoins were traded at a price of 5 cents each. The value of a single coin has since multiplied over a million times to over $70,000. Within a couple of years of its founding, Mt. Gox was responsible for 70 per cent of all bitcoin transactions in the world. In its heyday, Mt. Gox wasn't *in* the bitcoin market, Mt. Gox *was* the bitcoin market.

In 2014, the exchange suffered a massive hack: 850,000 bitcoins, or 7 per cent of the entire supply of bitcoin in the world, vanished overnight. At the time, they were worth around half a billion dollars; at today's exchange rate the haul would be worth over $60 billion. In what has been called the first bank run of the crypto era, 24,000 people lost their deposits and Mt. Gox declared bankruptcy. The exchange was such a big player in the crypto space that the failure of this one company slashed the price of bitcoin by half. A prominent New York newspaper declared, not for the first or the last time, that 'the bitcoin dream is all but dead'. Mt. Gox may have perished but its memory is etched into the lexicon. The company's name is virtually synonymous with failure: to be 'goxxed' is to be screwed.

Who hacked Mt. Gox? For almost a decade no one knew. Tens of billions of dollars in crypto simply vanished and no one could figure out where it all went. Years went by. The value of the stolen loot grew and grew. The crypto fortune plundered from Mt. Gox would eventually become worth three times more than the top ten bank heists in all recorded history combined. This whodunit only added to bitcoin's mystique. If the identity of the currency's founder, Satoshi Nakamoto, is the first major mystery in bitcoin lore, the murky origins of the black swan event that led to the collapse of the world's largest crypto marketplace would become a close second.

Why couldn't anyone figure out what happened? Well, the technology underlying the blockchain was so new, and the state of the art moving so fast, that law enforcement just didn't know what was going on. 'No one knew what happened,' Michael Gronager, the founder and CEO of Chainalysis, tells me. 'You have this idea, at least everyone I spoke to had this idea, that we are playing around with crypto and law enforcement of course knows everything we are doing and have a much better understanding. And I realized they were clueless. They had no understanding of how crypto worked. How to trace money. How to do anything.'

At the time of the Mt. Gox hack, Michael was working for Kraken, a rival cryptocurrency exchange based in San Francisco that he had set up a few years earlier. In the very public spectacle of the hackers running circles around law enforcement, Michael sensed an opportunity. Mt. Gox was the first big crypto crime, but it certainly wouldn't be the last. The world needed a new type of company that could run forensics on blockchain transactions. So he left Kraken to launch Chainalysis, a startup dedicated to tracing movements on the blockchain. They are now known as the blockchain detectives.

Chainalysis was soon appointed the official investigator for the Mt. Gox case. With the new startup's help, law enforcement was finally able to crack the case in 2023, a decade after the hack. They traced it to the operators of BTC-e, a cryptocurrency exchange based in Russia. US authorities would eventually indict two Russian nationals for the crime, one of whom was caught while vacationing with his family at a beach resort in Greece. Born out of the Mt. Gox collapse, Chainalysis now operates in over sixty countries and solves all sorts of crypto-based crime, everything from money laundering to drug trafficking. The company is valued at over $8 billion.

Gronager, who was born in Denmark and moved from San Francisco to New York City to launch Chainalysis, told me that it was just a better home for his company. 'I was working between Copenhagen and San Francisco and trying to do that is a nine-hour time difference, and it just doesn't work that well,' he said. 'It just tears you apart. It's not easy.'

'There was a conversation I had with my co-founders and we were like where do we best position ourselves on the map given that we build what we're building? And we had some of these discussions early on and the mindset was that the public sector will be important for us, be it law enforcement, be it regulators, whatever. We also are building a core part of finance. And none of these two things live on the West Coast.

'We did get told by investors that you have to move everything to the West Coast at some point. And we were like: we don't. And today it's pretty obvious that there's so many companies running out of the East Coast. So it's changed. And it's been easier. And we were kind of part of the early wave of moving some of tech to the East Coast.'

3

There's a lot of noise out there. There are opinions. But those cut both ways. And then there are any number of consulting type analyses which show professional-looking graphs pointing in this and that direction. Venture investing has gone down. Billion-dollar companies have gone up. Returns have gone sideways. Those sorts of things. But too often those glossy expositions confuse short-term changes with long-term shifts. Arrows that are pointing down today can just as easily be pointing in the other direction tomorrow.

So we're still left with the question: what's really happening out there? Is the Valley done? Or are its best days yet to come? It's hard to read the situation. I thought I'd try and get a lesson in how to interpret the Valley from someone who helped write its story. So I reached out to John Hennessy, the Chairman of Alphabet, the parent company of Google, to ask him how he saw what was going on and what might happen next.

Hennessy has spent more time on the inside of the inside of the Valley than perhaps anyone who's still active on the scene today.

He was only 25 years old when he first arrived in Palo Alto in 1977 to join the electrical engineering department at Stanford. It would prove to be a fortuitous move. Apple had launched a year earlier, Microsoft the year before that. Both companies were yet to launch a major product. All the big tech companies – IBM, Xerox, Polaroid – were still out on the East Coast. But change came fast. 'Quite frankly, I'm surprised at how once things began to swing how quickly it happened,' Hennessy tells me. 'Which was really the late 70s when Apple was founded, though by 89–90 the momentum had switched.'

Hennessy experienced the region's ascent mirrored in his own career. Rising swiftly through the ranks, he became the Chair of the Department of Computer Science in 1994 while still in his thirties, then the Dean of the School of Engineering in 1996, before succeeding Condoleezza Rice as Provost in 1999. And in 2000, at the age of 48, he was appointed the tenth President of Stanford.

He would go on to enjoy an unusually long tenure at the helm, sixteen years, from 2000 to 2016, a period which coincided with perhaps the most consequential phase of the Valley's history: the birth and boom of the dot-com era. As a faculty member he had seen the region take shape. As president he saw it go through its golden era as technologies developed at or near his campus reshaped economies and societies in virtually all parts of the globe. Google, Instagram, Snapchat, all came up during that time.

So how does someone who didn't just see it all happen but in a very big way made it all happen, assess what's going on now? I asked Hennessy if he thought the Valley was in decline. My query was met with blank stares. It was as if the man spends literally none of his time entertaining doubts about the future of the place he calls home. 'In the end it's the killer app that has driven the Valley time and time again,' he said. Small and medium-sized advances have always come from lots of different places. But the Valley's the place that consistently throws up those big, hairy, epoch-defining technologies that come up every so often and change everything: semiconductors, personal computers, software, smartphones. And it doesn't look like that's about to change anytime soon.

So what's the next killer app? For Hennessy, it's artificial intelligence, with the Bay Area firmly in the driver's seat. 'It is the centre of the next generation of AI companies, there'll be ones elsewhere, but certainly the Valley probably outnumbers anybody else by a factor of, I don't know, two, three, four or five probably?' And this next big wave may just be bigger than all the ones we've seen before. 'This technology is moving faster than anything I've ever seen, faster than microprocessors, faster than personal computers, faster than the internet, faster than web, faster than email.'

The world is moving from the dot-com era to the dot-ai era and the Valley is the place where that shift is happening. The most consequential companies of this next big wave are here. It could be argued that far from fading away the Valley is in fact going through a significant expansion. Most of the iconic names that we associate with the Valley have traditionally come from the southern part of the Bay Area. Apple? Cupertino. Google? Mountainview. Facebook? Palo Alto. But now the action is moving further up north to San Francisco.

The city was until recently a relatively minor player in tech. 'Only ten of let's say fifty years of technology explosion was really San Francisco relevant at all,' says Keith Rabois. 'It was always kind of a misfit for technology versus the South Bay.' But now SF can credibly claim to be the capital of the AI revolution. Some of the most recognizable AI companies are tightly clustered in or around one neighbourhood, Haye's Valley, now sometimes called Cerebral Valley, for its high density of AI companies. OpenAI, Anthropic, and Databricks – the three most consequential AI companies in the US if not the entire world – are all within a two-mile radius. Y Combinator too moved its headquarters out from Mountainview to San Francisco's Pier 70 in the Dogpatch neighbourhood nearby, where its portfolio has in recent years skewed heavily in favour of AI companies.

4

Even in China I came across ready acknowledgement that when it comes to AI, the US is still the place to be. 'I think in AI in terms of the fundamental technology and the research, you'll see contributions from China, but I would say the US is still ahead and will continue to be ahead for many years to come,' Ya Qin Zhang told me.

Zhang is in many ways to the Beijing tech community what Hennessy is to the Valley, an uncontroversial father figure who is as at ease in a corporate boardroom as he is in an advanced research lab. Zhang presented a sobering assessment of where things stand between the US and China in AI. 'There are components that China will improve at but, overall, the US is clearly in the lead.'

Zhang says that leadership in AI rests on four elements: algorithms, data, compute and talent. China is doing well in the first two but is still playing catch up in the latter two.

Compute refers to processing power. AI applications are resource-intensive, much more so than traditional software, and require their own specialized hardware. In the race for AI, brute power matters a lot: the more compute power you have, the faster you can train and run your AI algorithms. And here the US has practically cornered the global market for compute. Its lead comes down to one company: Nvidia.

Nvidia, also based in the Bay Area, is by far the single biggest winner of the AI boom. The Santa Clara headquartered chipmaker is the infrastructure backbone of the AI movement. Simply put, the company makes the chips that are used most often in AI applications. The company has been around since 1993 but its ascent into superstardom is recent. Its chips, called Graphical Processing Units, or GPUs, were initially designed for high-end gaming. But it later turned out that they could also be repurposed to do the heavy lifting that goes into handling AI algorithms. The company's swift rise soon followed.

Four out of five GPUs sold for AI applications are manufactured by Nvidia. In published AI research, its chips are used nineteen times more than all other chips combined. And these processors don't come cheap. Nvidia's most high-end chip, the H200, can cost more than $40,000. And still the market can't get enough of them. Customers have had to wait months or sometimes even years to get their orders filled. In 2023, one company, Meta, spent over $10 billion in a single year to buy 350,000 Nvidia GPUs. One analyst remarked, 'There's a war going on in AI out there, and Nvidia is the only arms dealer.'

In a message I heard from one industry expert after another, it has proved near impossible for startups and established companies alike to compete with Nvidia. Many have tried. Amazon, Meta, Google and Microsoft are all now in the chips business. But so far no one has succeeded in making a dent in Nvidia's lead. Why? It's just really hard to make chips. Even if the science and engineering are well-known, there's a lot of tacit knowledge that goes into designing and manufacturing high-end chips which can be hard to replicate across organizations. Just because someone can get their hands on the recipe to make chips doesn't mean they also have the skills to pull it off.

And this has handed Nvidia, and by extension the US, enormous advantages over overseas competitors. The US has banned the export of Nvidia's flagship chip, the H200, and its less powerful sibling, the H100, to China. These constraints have made life harder for Chinese companies, but not impossible. Many have been able to skirt these regulations. Nvidia is prohibited from exporting physical chips to China. But Chinese companies can still access compute resources through cloud providers based in the US. Often, they just buy these chips through subsidiaries outside China and deploy them in other locations in Asia, like in Singapore or Vietnam.

There's also a strong incentive for Nvidia to look the other way. Before the export controls were put in place, China was responsible for up to a quarter of the company's business. That's a big market to lose. Nvidia has designed lower powered chips whose compute is just under the threshold set by US regulators for export controls to China. These lower powered chips can always be deployed in clusters to achieve the same effect as the higher end chips. All of which is to say that the US might control the tap for AI compute, but a cat and mouse game is under way to blunt this advantage.

But China is not the only country impacted by America's grip over the global supply of high-end chips. Other countries have also been feeling the crunch.

Edouard Bugnion was born in Neuchâtel, in the French-speaking part of Switzerland. But he knows the Valley just about as well as any Bay Area native. The Stanford graduate co-founded VMware, a cloud computing startup, in Palo Alto in 1998. Twenty-five years later the company was sold to Broadcom for $69 billion, the largest tech acquisition in the Valley ever. Bugnion has since moved back to Switzerland, where he teaches at the Swiss Federal Institute for Technology in Lausanne.

Bugnion is just as conversant with what's happening with tech in the Bay Area as he is with the goings on in Europe. I asked him how he compared the two, and whether he thought the relative importance of the Valley is on its way up or down.

'Oh, I think it's going up,' he said. 'I think in the AI revolution they control the cards and they're going to go after it Silicon Valley-style, which is go first fast and think later.

'The American companies have this unfair advantage of having, first of all, the compute capacity, which is right now an artificially limited resource, which is benefiting one company economically, which is problematic – this is Nvidia. A dead painter has arbitrarily high value because the painter's dead. And we understand that the sky's the limit. A GPU is a GPU. At the end of the day it should not cost what it costs today. US companies have this unfair advantage of compute capacity and the data, they've ingested data, and so they have an edge.'

With the shift to AI, the gap between the US and the rest of the world in new technologies, which has now been around for almost a century, and is the subject of much anguish in places like Europe, is set to skew even further. And, for Bugnion, not even China comes close to mounting an effective challenge.

'And part of it is because it's too strategic,' he continues. 'I think at some point this is where geopolitics comes into play. The geopolitical assumption that the US will simply let the free market play out for AI in terms of interactions between the US citizens and AI systems and let China control that? I think that is just too simplistic. It's too risky.'

5

The US has a definitive edge over China in AI talent as well. The reason is simple. China has a lot of people. But the Chinese market for talent is largely confined to its own national boundaries. China chooses its talent from a billion people; the US, from eight billion. And that just gives the US a massive leg-up in the competition.

Over half of all billion-dollar tech companies in the US were founded by people born outside the US. Two-thirds of the tech talent working in the Valley was born overseas. Over half the residents in the Bay Area speak a language other than English at home. In China, things are very different. In Beijing and Shanghai, by far the most international cities in the country, not even 1 per cent of the population is foreign born. The Chinese tech industry is still very Chinese and, after the recent decoupling with the US, becoming even more so. The US tech industry is the most global in the world.

The competitiveness of the US tech industry in relation to international rivals might have less to do with the minutiae of its industrial policy and more with the unique nature of American society which makes it easy for anyone born anywhere to move to the US and call themselves American. That's a hard advantage to beat. That's hardly an original point. Ronald Reagan captured the sentiment in a speech in 1989 when he said:

> You can go to live in France, but you cannot become a Frenchman. You can go to live in Germany or Turkey or Japan, but you cannot become a German, a Turk, or Japanese. But anyone, from any corner of the Earth, can come to live in America and become an American.
>
> This, I believe, is one of the most important sources of America's greatness. We lead the world because, unique among nations, we draw our people – our strength – from every country and every corner of the world. And by doing so we continuously renew and enrich our nation.
>
> While other countries cling to the stale past, here in America we breathe life into dreams. We create the future, and the world follows us into tomorrow. Thanks to each wave of new arrivals to this land of opportunity, we're a nation forever young, forever bursting with energy and new ideas, and always on the cutting edge, always leading the world to the next frontier. This quality is vital to our future as a nation. If we ever closed the door to new Americans, our leadership in the world would soon be lost.

A major reason why American leadership in technology went uncontested for so long is because anyone who could challenge that supremacy chose instead to be a part of it. Some of America's most iconic technology companies were built by immigrants: Facebook, SpaceX, eBay, Yahoo and Sun Microsystems all had at least one immigrant co-founder. Other giants like Microsoft, Google and Uber are led by CEOs who are immigrants. When Stripe went public in 2016 at a valuation of $9.2 billion it made its 27-year-old co-founder Patrick Collison the youngest self-made billionaire in the world. Collison, who was born in Ireland, told the *New York Times* that he could not imagine building a company without immigrants: 'In the early stages of a startup you usually have a very specific set of things you need to do, and there's a very short list of people who are able to do them,' he said. 'The fact that the talent is here or that we can bring the talent here, that's what makes the whole thing work.'

For much of this international talent, it's the US university system that serves as their gateway to becoming American. In 2023, there were a million foreign students studying in the US, the largest cohort of international students globally. As discussed in the previous chapter, Chinese universities have been improving their position relative to the US in several rankings. But their progress rests on relatively narrow measures of research output. The culture of US universities is still aspirational for the best of the best in the world and no one, not China, and not Europe, come anywhere close. 'I think one of the struggles in Europe has been the balance between excellence and equity,' says John Hennessy. 'If you say all our universities are equal, what you've just said is you're not going to have any world-class universities because you can't, no country, not even the US, can afford to have all world-class universities.'

To be sure, Chinese companies have been stepping up their game in luring top global talent. Until just a few years ago Baidu had assembled some of the best AI talent in the world in its research division. The team was headed by chief scientist Andrew Ng, a Stanford professor who is considered one of the world's top authorities in AI. Soon other bold-faced names from the AI community followed. 'I mean you look at the crew that he had at Baidu and these people are like proper gangsters basically,' Nathan Benaich, the author of the 'State of AI Report', told me. This group at Baidu was responsible for Deep Speech 2, an AI program which used machine learning to understand human speech, still considered a major breakthrough in AI research.

But if restrictions on access to high-performance chips is one way the US–China competition in AI is playing out, making it harder for China to recruit top tier talent is another. International researchers have become more wary of working for Chinese companies, in part due to concerns about being seen as somehow suspect by US authorities. Many of the people who I spoke to who had worked in prominent positions in Chinese firms were often hesitant to talk about their experiences. Some tried to distance themselves or downplay the nature of their involvement, framing it as a minor detour in their careers, sort of like a professional summer abroad, rather than something that defined their work in any meaningful way.

The Chinese government runs over 200 talent recruitment programmes, the most high profile of which is the Thousand Talents Program, or TTP, launched in 2008 to bring high calibre scientists and researchers, drawn mainly from the Chinese diaspora abroad, back to the mainland. It has brought over 7,000 researchers back to China. But the US soon countered with the China Initiative, a Department of Justice programme launched in 2018 to prosecute Chinese espionage in American universities and research centres, which it alleged were taking place under the cloak of government sponsored programmes like the TTP.

The FBI opened over 2,000 cases under the China Initiative, and at one point a new case was being opened every twelve hours. This included the high-profile arrest and then subsequent dropping of charges against Gang Chen, a named Professor of Mechanical Engineering at MIT. The China Initiative, which came under heavy criticism from civil liberties groups for racially profiling Chinese Americans, has since been scrapped. But the environment of suspicion around people with close ties to China remains and few top-flight researchers, of Chinese origin or otherwise, are willing to risk the occupational hazards of working for Chinese companies and research institutions.

The biggest names behind the Deep Speech 2 breakthrough at Baidu, including Andrew Ng, would within a few years abruptly leave the company. They can now be found in the leadership ranks of the most prominent American AI companies. Dario Amodei co-founded Anthropic, the most significant US-based competitor to OpenAI; Adam Coates is the Director of AI at Apple; and a handful of others occupy senior positions in AI divisions at Nvidia and Google. Nearly all of them are based in the Bay Area.

So even if premature obituaries are once again being written about the Valley, the fact is that the Bay Area is still the centre of all the action with it comes to the next generation of AI hardware and software companies. This extends America's commanding lead in tech over the rest of the world, and that very much includes China. This asymmetry can at times be hard to grasp. Just six American tech companies – Microsoft, Apple, Alphabet, Meta, Amazon and Nvidia – are collectively worth more than all the companies listed on all the stock markets in all of mainland China combined. In terms of scale there's simply no comparison.

'Look, the artists go to LA, finance guys go to New York, tech guys go to Silicon Valley,' Bilal Zuberi, Venture Partner at Lux Capital, told me.

'I'm long on Silicon Valley. That takes nothing away from the rest of the world. But if I want excellence, I want excellence. It's just different. It's not just the specifics of what you do. It's the ecosystem around you. It's everyone. It's the diffusion that happens, that knowledge, that learning from each other, the support you can give to each other, the network that you build around each other and which then supports each other. All of that is really important.

'Ten years ago, I would've said for a certain number of people, China was emerging as something that was really interesting and powerful, but then China went in a completely different direction. The point is that if you have the opportunity to be surrounded by the best and surrounded by those that are learning the fastest and those that are able to execute the fastest so that you can see the mistakes they make and learn from them, but also the successes they have and learn from them, I think there's no better place for that than America.'

6

In *A Study of History*, the magisterial twelve-volume history of civilizations, Arnold Toynbee notes that 'growth takes place whenever a challenge evokes a successful response that, in turn, evokes a further and different challenge'. It would be a mistake to assume a linear progression of events in the US in the face of a changing world. It is for the first time since the end of the Cold War that America's sense of priority over the world is being seriously challenged. History has not been kind to those who have made overly hasty predictions of American decline without considering the country's remarkable capacity to adjust to change.

Many of the anxieties around the US and Silicon Valley's place in the world are a replay of debates that have already been had before. John Hennessy came of age during the Cold War amid widespread anxieties that the US was falling behind in key technologies. His father, an engineer who worked in the aerospace industry, encouraged him to go into computing after the Soviets fired the opening shot in the space race with Sputnik. And yet within two decades he would see the centre of gravity in technology shifting decisively first to a single country, then a single region, and then to a small patch of land centred on his university radiating outwards.

The scepticism around whether the Valley still has the magic or if it's all just smoke and mirrors, hype and marketing, is also something he has lived through before. Once, while still a junior faculty member at Stanford, he had managed to raise over $1 million for a research project, a huge sum of money for that time. He was presenting his work on a panel when the moderator asked the others for suggestions for what the young researcher should do with all the money he had managed to raise. One sceptic suggested that the whole thing was a fraud, it would never work, and Hennessy was better off just taking the money and disappearing to Barbados.

Hennessy did not in fact decamp to Barbados. And he did build out that technology. As a computer scientist he is best known today for developing the Reduced Instruction Set Computer, or RISC architecture, an approach to semiconductor design that made microchips faster and more efficient. It is now used in 99 per cent of all new computer chips. Hennesy's early work on the RISC would eventually earn him the Turing Award.

Champions build other champions. Fred Terman, a former provost of Stanford, who we'll talk about in a bit, is the one person most responsible for building Stanford into what it is today. But a less well-known fact is that he also spearheaded the development of the Korea Advanced Institute of Science and Technology in South Korea. The school, popularly known as KAIST, is now a top engineering school in Asia. A cynical way to look at this would be to see this as Terman seeding competition for his own school abroad. The more collegial view would be that one person built two great schools, with the second doing nothing to erode the standing of the first, and with both now coexisting in a symbiotic relationship.

In a similar vein, in my interaction with Hennessy, I could sense not an insecurity or a fear that the Bay Area's crown was now up for grabs but, befitting his stature as something of an elder statesman of tech, a certain noblesse oblige that seemed to say that he would love to see more places be able to do what they had managed to achieve in the Valley. But it's just not happening. Others just can't keep up.

'It's unclear that anybody, any of these other places will be a broad competitor across the entire IT space the way the Valley is,' he said. 'Maybe. But it hasn't happened. Attempts to do it in the past have been not very successful.

'The large players would like to have an innovation centre outside the Valley that has the same verve that the Valley has. They would love it because it would get away from some of the infrastructure and cost issues in the Valley. How might that work? We'll see, we'll see.'

Why the Valley?

Most people have some sense that the Bay Area is unrivalled in its place in the technology world. But the magnitude of its lead over others can be hard to grasp. Silicon Valley has a per capita GDP higher than almost every nation. It is in the same league as petro-states like Qatar and outpaces small enclaves for the ultra-wealthy like Macau and Luxembourg.

The region has an estimated GDP of almost a trillion dollars. If it was a country, it would be among the twenty largest economies in the world. It is home to 200 of the 500 largest companies in the world by revenue, seven of the top twenty largest companies in the world by market cap, and more than half of all American billion-dollar tech companies. Nearly half of all venture dollars invested in the US are invested in this region. By most measures – number of tech companies, valuations, dollars invested, salaries – the Valley matches or even eclipses the rest of the US combined.

How did this all come about? In the more usual telling of this tale, the history of the Valley begins in the middle of the twentieth century, with the arrival of silicon in Silicon Valley. But arguably its origins stretch much further back in the past, well into the previous century.

The story of the Valley can very plausibly be said to have started in the late 1800s, not in Palo Alto or even the San Francisco Bay Area but 7,000 miles away in the Bosphorus Strait, the narrow waterway that splits the city of Istanbul in two and marks the geographic separation between Europe and Asia and the symbolic divide between East and West.

In the winter of 1883, Leland Stanford Junior, the 15-year-old only child of an affluent West Coast couple went on a voyage through Europe with his family; a customary rite of passage for young men of means coming of age at the time. After spending Christmas in Vienna, the Stanfords headed to Constantinople, where young Leland was confronted with what seemed to him to be a strange and unfamiliar culture. 'When we arrived we thought we were in the strangest country we had ever been in before,' he wrote to his friend Lizzie Hull back home. 'No two Turks seem to be dressed alike because their clothes are of so many different colors ... We saw diamonds literally by the bushel and one emerald as large as your hand.'

Leland, a child of the Gilded Age, was born into a family of silver and servants after eighteen years of a childless marriage and was used to having attention showered on him and getting his way with his parents. His birth was announced at a lavish dinner party in Sacramento where after the guests were seated the waiter brought in a large silver platter with a cover and placed it in the centre of the table. The child's father rose to his feet to announce that there was someone he wished to introduce. The cover was lifted and lying underneath on a bed of blossoms was little baby Leland Junior. The child was promptly carried around the table and shown to each guest. 'He was smiling and went through his introduction very nicely,' Bertha Berner, Leland's mother's long-time secretary, wrote in her memoir in 1934.

So on this particular voyage – while cruising on the Bosphorus Strait – when young Leland insisted on taking the helm of the party's small steamboat his parent's couldn't help but let their son have his wish. 'He stood at the wheel all day long, with a sharp wind blowing in his face and spray dashing over the deck, for it was a rough day,' writes Bertha Berner. 'He was greatly excited and very happy.'

Later that evening, Jane thought her son looked a little pale. When they arrived in Athens the weather had taken a sharp turn and the snow was knee-deep. Leland was undeterred and went around visiting ancient temples and even snagged a meeting with Heinrich Schliemann, the acclaimed archaeologist who excavated the site of ancient Troy. But by the time the family reached Italy in February, the boy's health was in steep decline. 'The climate of Rome plainly did not agree with Leland,' writes Bertha.

Leland Junior had contracted typhoid and was soon convulsed with high fever, severe headaches and nausea. Antibiotics had not yet been invented and doctors resorted to wrapping the boy's body in ice-cold wet sheets. His hapless parents frantically rushed him first to Naples, then to Rome, and finally to Florence. 'For three weeks, alternate hope and fear reigned in the darkened room,' Herbert Nash, the boy's tutor, recounts in a short biography about his pupil. As Leland Junior's fever spiked and ebbed 'his mind was lucid at times, and at times wandering'. On 13 March 1884, two months before his sixteenth birthday, Leland Junior died in a hotel in Florence. Leland's father, who had stayed at his son's bedside the whole time, fell into troubled sleep on the morning of the boy's death. When he woke up he turned to his wife and said, 'The children of California shall be our children.' These words, according to the institution's own official history, were the beginning of Stanford University.

Junior's father, Amasa Leland Stanford, was in many ways the original Palo Alto tech mogul. Many of the Valley's current crop of tech elite made their fortunes by building networks and connecting people. That's just about how Stanford made his riches too. He was an early pioneer of one of the most advanced technologies to come out of the late nineteenth century: railroads. Stanford was one of the four people responsible for building the Transcontinental Railroad that connected the two ends of the US. Before Stanford came along, the 2,000-mile journey between the East and West Coasts took a perilous six months by horse-drawn carriages. He shortened that journey to only four days by train. Stanford was rewarded handsomely for ushering in this revolution in transport, making him one of the richest men in the world, a billionaire in today's terms. This wealth bankrolled his political ambitions, setting him up to become the eighth Governor of the State of California and then an elected representative to the US Senate, where he served until he died in 1893.

A year after their son's death, Leland and Jane Stanford founded the Leland Stanford Junior University in his memorial. This is still the school's official name, one that appears on the university's seal, is printed on every single diploma, and is the name under which the school still files its taxes every year even to this day.

The Stanfords devoted most of their personal wealth, a fortune of over $45 million, to the new university. This would amount to a billion dollars in today's terms. The grant included 8,180 acres of vast farmlands in Palto Alto, a donation so large that it makes Stanford University even today, 150 years after that original land allotment, one of the five largest private colleges in the US and the largest landowner in the Santa Clara Valley. The land given to Stanford was a third the size of the city of San Francisco and two thirds the size of the island of Manhattan. The value of this land, none of which has ever been sold, now tops $20 billion. 'Perhaps the greatest sum ever given by an individual for any purpose is the gift of Senator Stanford,' wrote Andrew Carnegie of the grant in 1889. Stanford is unusual among universities in that it owns not just academic buildings and sports facilities but also a shopping mall, a golf course and over 700 single-family homes.

Leland Stanford was a visionary. But even he could not have foreseen what his university and its surroundings would become. When he peered into the region's future, he saw not Apple but actual apples. He thought the future of the Valley lay in exporting fruit. 'Some day you will see Palo Alto blooming with nearly all the flowers of earth, and the fruit and shade trees of every zone,' he wrote in 1887. 'In the future we shall can this fruit and send it all over the globe in exchange for wealth, which shall build us monuments of art and bestow upon us those luxuries which God has intended we should enjoy.' And that is precisely what happened. Long before this was Silicon Valley, it was, in the words of a local poet, 'the valley of heart's delight', mile after mile of fruit farms and orchards that stretched in every direction. By the turn of the twentieth century this region was the world's foremost exporter of cherries, pears, apricots and prunes; up to a third of the world's supply of these fruits came from this patch of land. That agrarian legacy lives on. Even today the Stanford campus is known to its students as 'the farm'.

Leland Stanford might not have seen where the university would go in the future but he was surprisingly prescient about how long it would take to get there. In 1887, he wrote: 'A university, like a tree, is planted in the soil to grow at first unseen. I shall hope for a natural process. It shall not be my fault if the growth of the university be not slow, gradual, and steady.' Amasa Leland Stanford had set in motion events that would change the fate of not just this region but the entire world. But it would take another generation before the institution he created would produce someone who would set it on course to become what we know it to be today.

2

Before Leland Stanford there was nothing. After him there was something, a university, but not a very good one. Stanford was for at least half a century a land-rich but money-poor college which barely mattered on the national stage and almost went bankrupt within the first decade of its founding. 'There is about as much need for a new university in California as for an asylum for decayed sea captains in Switzerland,' the *New York Mail and Express* sneered in 1891. Leland Stanford was the person who got things going, but he isn't the one who took the Valley or even the university that is his namesake to the stratospheric heights they occupy today. That distinction belongs to Frederick Terman.

Fred Terman was born in 1900, seven years after Leland Stanford passed away. Terman's entire life was defined by Stanford and he in turn shaped the school more than any other single individual. Terman was practically born into it: his father was a prominent psychologist who developed some of the earliest IQ tests and served as the chair of Stanford's psychology department for over twenty years. Terman spent his early childhood hunting rabbits and looking for butterflies on campus grounds. He got both his undergraduate and graduate degrees there and soon after joined the faculty in 1925 which is where he stayed until his retirement in 1965.

An engineer by training and administrator by inclination, Terman rose through the ranks to become the dean of engineering and later provost and vice president of the university. At Stanford, which generally lacked funding and prestige when he first arrived, he envisioned assembling a 'community of technical scholars': a modern version of medieval societies of learning in old European centres like Heidelberg and Oxford where there could be, in his words, 'a continuous ferment of new ideas and stimulating new challenges'.

The phrase that is most associated with him though is 'steeples of excellence'; that was his city on a shining hill, his *Veni, vidi, vici*. It summed up his oft-repeated conviction that if Stanford was to be a world-class university it would have to steer clear of trying to be good at everything and focus instead on being the best at a few important things. 'It's better to have one seven-foot jumper on your team than any number of six-foot jumpers,' he wrote. And that for him meant building out the school's engineering and science departments.

Like the founding of Stanford, the events that shaped Terman's thinking were also the unintended consequences of tragedies that unfolded closer to shores of Europe than the orchards of Palo Alto. This improbable story starts with the sinking of the *Titanic*. Just before midnight on 14 April 1912, when the vessel hit an iceberg a thousand miles from the coast of New York City, there was another ship sailing only 20 miles north of the disaster. The SS *Californian*, a British steamship, was so close to the *Titanic* that later in the night its occupants could see dimly in the distance the ship's stern lift vertically out of the water and its hull snap in half.

The *Titanic* sent distress signals to the Californian almost as soon as it hit the iceberg at 11.40 p.m. But the ship's wireless operator had just minutes earlier switched off his equipment and gone to bed. Over a thousand lives were lost because help could not arrive in time. The avoidable catastrophe prompted Congress to pass laws to revamp maritime communications and made it mandatory for all ships to carry advanced radio equipment. A boom in the radio industry soon followed. As a port city, San Francisco and the wider Bay Area were direct beneficiaries of this new wave of interest in wireless technologies.

The region soon became the unlikely capital of the radio revolution. 'What Edison's Menlo Park is to the incandescent lamp, Palo Alto is to radio and the electronic arts,' wrote the *Palo Alto Times*. The Federal Telegraph Company, or FTC, based at a single-family house on 913 Emerson Street, became one of the most prominent names in the industry. It was back then to radio what Tesla is to electric cars or Apple is to smartphones. It was at the FTC that the vacuum tube amplifier and the oscillator were invented, technologies that had applications in not just radio but also television and all other electronics. An entire culture of amateur radio hobbyists and tinkerers sprang up around the company. Fred Terman, still in his teenage years, was taken by the looming presence of 50-foot poles strung with miles of aluminum wires in his neighbourhood. He managed to land a summer job at the FTC which sparked a lifelong interest in radio engineering. It would eventually become the subject of his doctorate at MIT.

Terman's advisor at MIT was Vannevar Bush, an influential science administrator who was later tapped by President Roosevelt to head the Office of Scientific Research and Development, or OSRD, set up in June 1941 with the purpose of bringing science to bear on the conduct of warfare. Six months later the Japanese bombed Pearl Harbor and the US formally entered World War II. The OSRD became the centre of wartime R&D efforts. It was here, under Vannevar Bush's direct supervision, that the Manhattan Project was launched. Bush mobilized 6,000 scientists to help with the war effort, including his former student, Fred Terman. Terman was by now heading the electrical engineering department at Stanford and was considered the world's foremost authority on radio engineering, the man who practically wrote the most widely used textbook on the subject.

Bush needed Terman's help with one specific problem. The cornerstone of the Allied military strategy in Europe was a strategic aerial bombing campaign. US and British aircraft took off from England and flew for seven hours over hostile enemy territory to drop bombs on critical infrastructure in Germany. But in the early phases of the war they were thwarted by a sophisticated German aerial defence system. The Germans had hundreds of early warning radar systems and thousands of radio-controlled anti-aircraft guns scattered across occupied Europe. These defences were taking down up to a fifth of all Allied aircraft and killed, wounded, or captured over 160,000 US and British airmen. Bush wanted Terman to figure out how to neutralize Germany's air defences.

Terman became the head of the secret Radio Research Lab (RRL) at Harvard in 1942. The entire purpose of the lab was to develop effective counter-measures to German air defence capabilities. Terman built the RRL up from just one person, himself, into a formidable 850-person research organization in just two years. The lab effectively turned the air war into a form of sophisticated electronic warfare. It was here that engineers developed radar jammers and came up with a technique to airdrop tin foil over German air defence systems to saturate their detectors with noise. By the end of the war, the US was flying thousands of planes at a time in formation equipped with radar jammers developed at the RRL that were dropping up to three-quarters of all aluminum manufactured in the US as chaff over German occupied territories. The Allies were able wrest control of the skies over Europe and win the war as a direct result of Terman's work at the RRL.

3

Terman came out of the war with two conclusions. The first was that even if he had a sentimental attachment to Stanford, the East Coast establishment still saw it as a second-rate institution. Why else would Bush make him move to Harvard to head up a radio lab that could just as easily have been based at Stanford? In 1943, while the war was still in full swing, Terman wrote a letter to a colleague in which he laid out his conviction that Stanford faced a stark choice: it could either take active measures to ascend the ladder of academic respectability and be taken seriously as an institution, like Harvard, or keep doing what it was doing and become even more comfortable with its own mediocrity. 'The years after the war are going to be very important and also very critical ones for Stanford,' he wrote. 'I believe that we will either consolidate our potential strength and create a foundation for a position in the west somewhat analogous to that of Harvard in the east or we will drop to the level somewhat similar to that of Dartmouth, a well thought of institution having about two per cent as much influence on national life as Harvard.'

His second conclusion was that ascending said ladder of respectability would require the school to forge a closer relationship with the national security community. During the war the OSRD spent nearly $450 million on weapons R&D. MIT got $117 million, Caltech $83 million, and Harvard and Columbia $30 million: money that proved to be practically transformative for these institutions. Stanford got only $50,000, further confirmation of its diminished standing in the eyes of the broader academic establishment. Terman wanted a seat at the big table and knew that the only way to get there was to tap into this vast pool of defence funding. Soon after he returned to campus in 1945, Stanford opened a lobbying office in Washington DC.

It found in the government a willing partner. At the end of World War II it was clear to US policymakers that the country's scientists had been at least as responsible for delivering victory as its soldiers. The paradigm had shifted from industrial warfare to technology-driven conflict. In an address to Congress given exactly a month after the US dropped the atomic bombs on Japan, President Harry Truman declared: 'No nation can maintain a position of leadership in the world of today unless it develops to the full its scientific and technological resources . . . no government adequately meets its responsibilities unless it generously and intelligently supports and encourages the work of science in university, industry, and its own laboratories.'

Vannevar Bush, now the closest thing to a nerd-hero-warrior that the country had ever seen, made an appearance on the cover of *Time* magazine in 1944. The following year he wrote an influential report titled 'Science, the Endless Frontier' which argued that 'the research scientists of the country must be called upon to continue in peacetime some substantial portion of those types of contribution to national security which they have made so effectively during the stress of the present war'. The war was over. But war-like spending on defence R&D would continue. And thus began the military–industrial–academic complex.

Stanford became one of the capitals of this complex. Terman soon recruited top researchers who had worked with him at the Radio Research Lab at Harvard to join him at the newly formed Applied Electronics Lab at Stanford, effectively doubling the size of the university's engineering programme overnight. The Office of Naval Research wrote them their first cheque to fund basic research on microwaves. By 1955, over half of the staff working in Terman's lab held security clearances for classified projects. Students were writing their masters and PhD theses that were classified. Terman became an advisor to practically every branch of the US military: army, navy and air force. Within fifteen years Stanford went from getting close to no military funding to ranking behind only Harvard and MIT as the biggest recipient of federal research dollars, cash that catapulted it into the academic elite.

Building steeples of excellence in science and engineering was the first element of Terman's plan to remake Stanford. Getting defence dollars into the system was the second. The third was to convince talented engineers coming out of the university to start their own companies near campus instead of going off to work for established companies out on the East Coast.

The most famous of these were William Hewlett and David Packard. Their company, Hewlett-Packard, founded in 1938, practically created the template of a scrappy tech startup launched by smart twenty-somethings in a garage which then disrupts entire industries on its way to becoming one of the largest companies in the world; the basic schema that animates Valley companies even today.

Terman also built Stanford Industrial Park on over 600 acres of campus land. The idea was to make it easier for engineers to start their own companies. This was another home run. One of the facility's early tenants was William Shockley, a brilliant engineer who invented the transistor, the basis of all modern electronics. Shockley won the Nobel Prize for this achievement and his company, Shockley Semiconductor Laboratory, was an early pioneer in the industry. Shockley was a genius but also an outrageously bad manager and, as it later turned out, an unrepentant racist, and so in 1957 eight of his best engineers broke off to form Fairchild, a rival semiconductor company. This act of apostasy by what came to be known as the 'traitorous eight' has over time become the central plotline of Valley lore symbolizing a rejection of hierarchy, authority and orthodoxy: capturing in a single act of defiance the essence of the Valley's self-perception of what it's all about.

Fairchild's timing was fortuitous. Two weeks after the company was founded, the Soviets launched Sputnik in what came to be seen as the opening shot of the space race. A torrent of money started flowing through research labs again. The National Aeronautics and Space Administration (NASA) and Defense Advanced Research Projects Agency (DARPA) were formed. Up to 80 per cent of government spending on R&D went to weapons research. Fairchild was a direct beneficiary of this largesse. The first batch of 100 transistors manufactured by the company were used to build the computer in the B-70 bomber. America's response to Sputnik, the Apollo programme, was also powered by Fairchild transistors. In 1964 alone, NASA bought 100,000 integrated circuits from the company. Four out of every five dollars the company made were made from government agencies. This was crucial to the development of not just Fairchild as a company but also integrated circuits and by extension semiconductors as a technology. In the early years, Fairchild's products were too expensive for private customers. It was government demand that kept the company afloat until it could improve its unit economics enough for its wares to be affordable to private companies. Mass adoption soon followed, eventually catalysing the digital revolution.

Fairchild was patient zero. It seeded the distinctive startup culture that eventually dug its roots deep into the Valley. Its management practices became the norm of the Valley way of doing business: informality, flat hierarchies, venture money and equity-based compensation for employees. But its most enduring contribution was in starting the virtuous cycle of success breeding more success. Fairchild's alums, known as the Fairchildren, went on to found some of the most iconic names in tech. 'Every time we came up with a new idea, we spawned two or three companies trying to exploit it,' wrote Gordon Moore, Fairchild alum and founder of Intel. An astonishing 70 per cent of publicly traded tech companies in the Valley trace their lineage back to this one company. The very name 'Silicon Valley' was coined by a journalist writing about Fairchild and its influence on the semiconductor industry in the weekly *Electronic News* in 1971.

'Here's to the crazy ones, the misfits, the rebels, the troublemakers, the round pegs in the square holes, the ones who see things differently, they're not fond of rules,' began Steve Jobs in that iconic Apple commercial from the 1990s. But this version of the Valley's self-portrait as a place where people just think differently, the artist in a turtleneck holding his chin, obscures the other half of the story: that it was shaped just as much by those who made the rules as those who chafed against them. 'It isn't a story simply of an entrepreneurial state or entrepreneurial hustlers in their garages. It's both,' Margaret O'Mara, Professor at the University of Washington and author of *The Code: Silicon Valley and the Remaking of America*, told me. 'And that is something that, particularly in an American context where we are very binary about these things, it is very difficult for people to recognize that these two things operate in this very symbiotic fashion.' Far from being fenced off in its own little corner on the outer reaches of the Western seaboard away from the embrace of American history, the history of the Valley *is* American history, wrapped up as it was in the geopolitics of World War II and the Cold War. Hidden under the hood of the Valley's shiny entrepreneurial sports car, O'Mara points out, was a big government engine that ran on federal dollars.

4

A lot of universities get defence dollars but not all of them seed thriving entrepreneurial ecosystems around them. Harvard and MIT got a lot more love from the defence community for a lot longer than Stanford. And even if Boston has done well, the Bay Area has clearly left it behind. So there's more than just universities and government contracts at play. The question remains: why the Valley?

AnnaLee Saxenian has spent much of her life thinking about this question. In the 1970s, just as the Bay Area was beginning to make a name for itself as a hub for semiconductor companies, Saxenian was a young graduate student in regional planning at the nearby University of Berkeley. She came across a news piece about how the booming electronics industry had made San Jose the fastest growing city in the US. So she decided to write her thesis about that.

Saxenian spent months interviewing semiconductor executives to try and figure out what was going on. In her graduate thesis, published in 1980, she concluded that growth was unsustainable and the boom wouldn't last. Her reasons would be familiar to those predicting the Valley's demise today: housing had become too expensive, traffic was too congested, salaries were too high, infrastructure was stretched to its limits.

Saxenian's assessment was influenced by the product lifecycle theory developed by the economist Raymond Vernon at Harvard. It posited that industrial clusters everywhere go through similar phases. First lots of companies are born. Then the bad ones die. Then the ones that remain consolidate into a few big firms. And then they relocate to lower cost regions. 'I argued that Silicon Valley would follow existing regions like Detroit or Pittsburgh,' Saxenian tells me.

Within a few years it became clear that the Valley had ignored her predictions and just kept on growing. So when Saxenian started her PhD at MIT she wanted to find out where she had gone wrong. In her graduate thesis she had concluded that Silicon Valley would be just like everyone else. In her doctoral dissertation, Saxenian asked: what makes this place so different?

Regional Advantage: Culture and Competition in Silicon Valley and Route 128, a book based on AnnaLee Saxenian's doctoral work, published in 1994, has become the standard explanation for why the Bay Area edged out other competitors to become the uncontested leader in new technologies. She compared the tech cluster that surrounds Stanford in the Bay Area to the one that came up around MIT and Harvard in Boston, called Route 128, to figure out why one boomed and the other faltered.

In the 1970s, it was Route 128 and not the Valley that was the leading tech cluster in the US. The region, hailed as the 'Massachusetts Miracle' and 'America's Technology Superhighway', was home to the most prominent tech companies of the time, like Vannevar Bush's own Raytheon, Polaroid and DEC. 'When you wanted to talk to the movers and shakers in the computer industry you got on a plane and you flew back to either New York or Boston, and that's just the way it was,' recalls John Hennessy.

Saxenian thinks that the difference in performance between the two regions is in part an unintended consequence of a minor kink in California's labour laws. California didn't enforce non-compete clauses as vigorously as other states, so employees were free to move around. In Boston, switching jobs was hard and it was not unusual for people to stay with the same employer all their lives. 'I knew people who worked for the minicomputer companies back east, and if you went from one company to another, you were considered a traitor, people shunned you, they wouldn't even talk to you,' recalls Hennessy. In the Bay Area, you could leave a company on Friday and go work for its rival the following Monday. People had technology loyalty, not company loyalty. They thought they worked for the Valley and not for a specific company. As people moved around so did new ideas. The free movement of information triggered a strong recombinatorial effect: ideas from distant domains could come together in unusual combinations to create new technological forms.

This bred two very distinct corporate cultures, one open and the other closed. East Coast companies were vertically integrated, autarkic systems where staff turnover was low, the value of secrecy was high and every effort was made to keep knowledge locked up within the walls of the organization. These firms were discrete, self-contained units with firmly etched boundaries. In the west things were a lot more fluid. Here companies functioned more like a web of interconnected nodes with porous membranes that seeped ideas freely into their environment. In the east there was a cluster, in the west there was a network, and that made all the difference.

Saxenian, who is now a Professor at Berkeley, where she has been on the faculty for over thirty-five years, thinks that this openness and flexibility has also made the Valley more adaptive to change. The region has enjoyed unusual longevity. The norm among industrial era clusters was that they emerged in tandem with a new technology and then declined and faded as that technology got replaced by the next or as it got taken up by international rivals. That's the story of Detroit, Dayton and Pittsburgh. But the Bay Area has been able to jump from one technology to the next and one industry to the next with relative ease.

The Bay Area too has weathered its share of storms. The semiconductor industry which gave birth to this region in the 1950s was largely outcompeted by Japanese rivals by the 1980s. In 1980, the top three semiconductor suppliers were American companies Texas Instruments, Motorola and Philips. By 1986, they had been replaced by Japanese firms NEC, Toshiba and Hitachi. 'The United States companies claim the Japanese success comes partly from the backing of their government, which had restricted American entry into the Japanese market and allowed the Japanese companies access to low-cost capital,' wrote the *New York Times* in 1982, a description that bears obvious parallels to anxieties about the challenge from China today.

'When I was writing, people would write in and say in the early nineties, those entrepreneurial firms in Silicon Valley, they'll disappear, they're going to go out of business,' Saxenian tells me. And yet four decades later it is Silicon Valley that is still standing and the Japanese semiconductor industry that has faded away. Other hubs emerge and then calcify. But the Valley's fluidity, everyone and everything always in flux and never settled, makes it regenerate over and over in the face of relentless change. Forty years of watching the Valley have made Saxenian wary of the naysayers. After all, she used to be one herself. 'There's just been a series of predictions of its decline,' she says. 'But I've learned not to bet against Silicon Valley because I already did that once and I was wrong.'

5

Like any historical question, it would be a stretch to say that there's anything close to a consensus among people who study the Bay Area about why it became the centre of new technologies. Sebastian Mallaby, author of *The Power Law: Venture Capital and the Making of the New Future*, thinks that mainstream accounts of the Valley's success often overstate the importance of a single individual, like

Fred Terman, or a single technology, like semiconductors, or a single institution, like Stanford, at the expense of a more systemic explanation which would place the relative emphasis on larger shifts in how businesses are funded and risk is approached. Mallaby says that the rise of venture capital, which turned conventional finance on its head, is central to the story of why it all happened and why it happened here.

'I was interested in the different theories that people presented to me about why Silicon Valley had grown up in Silicon Valley,' Mallaby tells me. 'Was it because of Stanford being there? Was it because of defence contracts? Was it because the weather was better on the West Coast? What was it? And I developed a strong view that what differentiated Silicon Valley from Boston – which had the early mover advantage in all this – was the nature of the risk capital, the fact that the VCs were just different than they were in Boston.'

These days it seems perfectly normal for twenty-somethings with no background in business to raise tens of millions of dollars and start a company. But just a few short decades ago this would have seemed like a bizarre way to do business. Back then, unless founders had collateral, years of experience in their industry, and long track record of operating a sensible business, which meant, typically, that they could show that their enterprise was, at a minimum, profitable, no sane institution would ever lend them money. But now that seems like a downright quaint approach to investing. How did that happen? The answer: venture capital.

Traditional investors assume that their investments will yield returns according to a normal, bell curve distribution. In this type of investing, risk is your enemy. You invest in ten businesses. Two do very well. Two do very badly. Six fall somewhere in the middle. The very good cancels out the very bad and you end up with a reasonable outcome overall. In venture investing, risk is your friend. You invest in ten highly risky ventures. Nine fail. But the one that wins, wins so big that the nine losses just don't matter in the cumulative.

This shift in the mindset of tech investors who went from chasing inverted U-shaped returns to L-shaped returns, an approach pioneered in Boston but which found its fullest expression in the Valley, changed the game for entrepreneurs. For most people the hardest part of starting a business is finding the money to make it happen. In the absence of investors willing to bet on unproven founders, entrepreneurs have to either risk their own savings or give up on their dreams. Venture capital de-risks entrepreneurship by allowing founders to take risks with other people's money. It also simultaneously democratizes it by making money available to those who otherwise wouldn't be able to start a business. This easy-access capital effectively acts as jet fuel for new venture creation.

Where a traditional investor's biggest fear is losing money on a bad investment, a venture investor's biggest fear is missing out on getting into a good one. As one VC explained to me: 'If I invest ten dollars in a company and it turns out to be the next Theranos, I just lose ten dollars. But if I don't invest ten dollars in a company and it turns out to be the next Google, I lose 100x or 1000x of my money. In venture the opportunity cost of missing out on a good thing is way higher than the cost of buying into a bad thing.'

This rejigs incentives for moneymen to shift their attention from an obsessive focus on the risks of an investment to an obsessive focus on its rewards. This culture of FOMO is stoked by stories of VCs missing out on life-changing opportunities. Bessemer, one of the firms that founded the industry, lists on its website its 'anti-portfolio', companies it had an opportunity to invest in, but didn't. It's a painful read. Apple, Airbnb, Google, Facebook, Intel, PayPal, Snap, Zoom, Tesla: practically trillions of dollars in forgone value. If one were to venture a guess, the value of their anti-portfolio probably eclipses their actual portfolio. 'If we had invested in any of these companies, we might not still be working,' notes the firm.

This is hardly a commentary on Bessemer's skill as an investor. The firm is still one of the best in the business. It only underscores how notoriously difficult it is to judge how game-changing an idea can be while it's still in its infancy. Even giants of the tech industry have had their misses. John Hennessy tells the story of how when he was the President of Stanford two students named Larry and Sergey came to him with the idea of a new search engine. He didn't see the point. Why does the world need another search engine when it already has one in AltaVista? Hennessy is now the chairman of that same company, Alphabet.

The fear of regret is a powerful motivator, and VCs have over time become progressively more liberal with doling out funds. Lots of experiments get tried, most fail, but the few that succeed turn the wheel of progress. This firehose of money chasing risky ideas speeds up the velocity of innovation. There are other ways in which the venture industry has de-risked entrepreneurship. When someone starts a company, they're not just risking their money, they're also risking their careers. What if, as is likely, things don't work out? Enter the benevolent venture capitalist. They can just reassign the unsuccessful founder to another company in their portfolio and, presto, the career risk has vanished.

Mallaby describes in his book the story of how John Doerr of Kleiner Perkins convinced a sceptical Eric Schmidt to join Google as its CEO in 2001. The startup was less than three years old. Schmidt had already been the CEO of Intuit, a public company. Google was a gamble. What if it failed? Doerr assured Schmidt that if things didn't work out he could always find a place for him in another one of his companies, which is what convinced Schmidt to take the leap. And the rest is history.

The venture industry deserves credit for many of the achievements that are usually laid at the altar of government. Like, for instance, the internet. The conventional story goes that DARPA created the internet. Wrong, says Mallaby. The internet was mostly a governmental and academic network until UUNET, a venture-backed company that was the world's first internet service provider, came along in 1987. By offering an internet connection to anyone willing to pay for it, UUNET removed the technology from the confines of government labs and made it the mass phenomena that we know it to be today.

The VC model didn't just change the Valley. It also changed China. Here too Mallaby challenges our conventional understanding that the rise of China can be attributed primarily to its government's competence at long-term planning. He says instead that the country's fortunes changed as a direct result of the arrival of American-style venture capital twenty years ago which is what gave rise to the country's tech powerhouses Baidu, Alibaba and Tencent. Investments were structured by American Silicon Valley lawyers, with dispute settlement under New York law, and these companies grew by issuing Silicon Valley-style stock options to early employees.

Venture capital changed the risk calculus for starting businesses on a systemic level. Mallaby calls it a machine for manufacturing courage. And for him it was this more than anything else that supercharged the innovation engine of the Valley and beyond. 'A willingness to bet on these entrepreneurs and their visions has made Silicon Valley the innovation engine of the world,' he says. 'Try and fail, don't fail to try.'

CHAPTER THREE

Busting Monasteries

'We're breaking it down from this being about the monasteries in Silicon Valley and in Boston and maybe one or two other places to a reformation, a renaissance back to Europe.'

I

When DeepMind was founded in a small ten-person office at the top of a townhouse next to the London Mathematical Society in Russell Square in the autumn of 2010, London was still an unlikely birthplace for a startup which with AlphaGo would within six short years produce one of the biggest leaps in the evolution of artificial intelligence. DeepMind was singular. Until it came along, London had produced few, if any, globally relevant tech companies for at least half a century. At the time of its founding there were only eight billion-dollar tech companies in the entire country. The hottest startups on the tech scene were companies like Unruly, Songkick and Dopplr: names that are still recalled with affection by those who knew them then but which never really gained much familiarity outside the UK and which have long since receded from popular memory.

DeepMind is of the same vintage as Uber, Twitter and Stripe.

But while its American peers raised billions in funding within a few years of their founding, the London-based AI upstart came up in relatively modest circumstances. In 2010, the ratio of venture money invested in Silicon Valley versus London was 7:1. DeepMind never really got much love from British investors. Few were willing to take a risk on an obscure startup working on sci-fi type technologies in a geographical context known more for high finance than high tech. DeepMind is now thought of as the icon of the UK tech scene. But the inconvenient truth is that most of the money that got it going came from overseas investors.

DeepMind was still little known outside the AI community when Google bought it for a reported $500 million in 2014. That one of the world's largest tech companies would part with a small fortune to buy a still unproven four-year-old British startup was seen as much needed validation of the UK's deep tech credentials: a win for DeepMind and a double win for London. But, some wondered, was it really a win for Google? At the time of its acquisition DeepMind had only seventy-five employees, no products and no revenues. Google had forked out nine figures for what was essentially a research lab masquerading as a company. DeepMind's commitment to science was admirable. But would it ever pay commercial dividends? The startup was sexy. But would it ever be able to show me the money?

A decade later, question marks about the financial prudence of the acquisition have turned to face the other way around. In hindsight, it is DeepMind and not Google whose dealmaking acumen is most frequently called into doubt. Did the startup sell itself too early and for too little? That certainly seems to be the sentiment gaining ground in tech circles in the UK: that a big American corporation made away with the crown jewel of British tech at bargain price. Google made $66 billion in 2014. The $500 million that it splashed on DeepMind seems in retrospect to be mere pocket change. It doesn't even rank among the top ten biggest acquisitions made by the Mountain View-based tech giant. And yet it may turn out to be its most consequential.

How much would DeepMind be worth today if it had stayed independent? Would it, like Anthropic and Databricks, be worth tens of billions? Or would it, like ByteDance and OpenAI, be worth hundreds of billions? To those inclined to still see things from a national perspective, a company like DeepMind is practically priceless. Not so much a business as it is a strategic asset, less Walmart and more Manhattan Project: among the best of a handful of research efforts globally that can crack the hard problems of the most important technology of our time. Ian Hogarth, the chair of the UK Government's AI Foundation Models Taskforce, has written: 'I find it hard to believe that the UK would not be better off were DeepMind still an independent company.'

These hypotheticals, which find much currency in Britain's commentariat, are as alluring as they are futile. Coulda woulda shoulda. It is what it is. What can you do about it? It might just be that with the Google offer DeepMind was presented with a fait accompli: not so much a real choice as an illusion of optionality. DeepMind took the blue pill. But was there ever really a red pill? It said yes and got that half of the decision tree. It could have said no and then done what exactly? The company struggled to find local sources for the relatively small amounts of money it needed to get going. Without Google where would it have found the gargantuan amounts of money it needed to grow? Humayun Sheikh, one of DeepMind's founding investors, has said that the startup would have 'probably failed' without Google. It needed a wealthy offshore patron that could underwrite the billions it would accrue in losses until it honed its unproven technology and, more importantly, provide it with the compute it needed to train and run its resource-intensive algorithms. DeepMind was enabled by the environment in which it came up. But it was also constrained by it.

2

London is now a different place from what it was fifteen years ago. Back then, DeepMind practically carried the entire country when it came to category-defining startups. Now there's an entire cohort of companies that have cropped up around it. And they're big. Checkout.com, Revolut and Global Switch are all worth tens of billions of dollars, multiples of what DeepMind sold for a decade earlier.

When Sonali De Rycker, a partner with Accel, first moved to London from the US in 2001, she didn't find much to get excited about: 'It was really like watching paint dry, everything was slow,' she recalls. In the intervening two decades things have changed: 'The level of activity that we see today is almost like it's a different business from when I first started,' she says. By most measures – size of the tech economy, number of billion-dollar companies, cumulative valuations of startups, volume of venture dollars invested – the UK is the third most important tech economy in the world, behind only the US and China. In Europe, it's the clear leader. In 2023, around $22 billion of venture funds were invested in UK startups, which is more than the next three countries on the list – Germany, France and Sweden – combined. Only three countries have produced more than a hundred billion-dollar tech startups in the past quarter century: the US, China and the UK. That's rarefied company.

Number three is impressive. It's a podium finish. It's probably the best outcome any country of this size could hope to attain given the sheer scale of the competition. But even if things in the UK's tech sector are better than they have ever been, and every single graph points up and to the right, and even if the country is, in a description that I heard repeated in every conversation I had in the UK, 'third globally and the best in Europe', the mood on the ground is far from triumphant.

The past casts a long shadow over Britain. Sentiment in the tech sector is hardly immune to wider anxieties about the country's relative decline that some say has been going on since 1945, others since 1870. In an article in *The Atlantic* titled 'How the UK became one of the poorest countries in Western Europe' the author Derek Thompson argues: 'The UK, the first nation to industrialize, was also the first to deindustrialize. Britain gave rise to the productivity revolution that changed the world, and now it has some of the worst productivity statistics of any major economy.'

Another reason why any self-congratulation over the UK's place among the top three tech economies tends to be somewhat muted is the sheer size of the gap between it and the other two contenders. In 2024, the total value of all British tech companies combined was about a trillion pounds. That's about a third of the market value of just one American tech company, Apple. The total market value of all the companies on the London Stock Exchange is less than the market value of one US tech company, Nvidia.

The situation in the wider European region is not much different. The combined value of all the tech companies in all of Europe is far less than the value of one company, Microsoft. The total value of every single company listed on every single stock exchange in every single country in all of continental Europe is less than the cumulative value of just five American tech companies: Nvidia, Apple, Microsoft, Meta and Alphabet. The Americans are not in a different league. They're playing a different game.

A trillion is an almost inconceivably large number, a gratuitous unit for measuring corporate value that would be comically out of place literally anywhere other than America. Consider: even if you spent a million dollars a day it would take you three thousand years to spend a trillion dollars. Even Chinese companies aren't spinning those sorts of numbers just yet. China has about a dozen tech companies that have crossed a $50 billion valuation. The UK? Only one. Third place is scarce consolation when the global tech order is essentially a two-party system. One way to look at the UK is that it's the best of the rest. The other, that it's the number one loser.

It is at first hard to see why the UK is a relative underperformer compared to the US and China. After all, the country does seem to have all the ingredients that one would imagine go into building a competitive tech sector. It's rich. An average person living in Britain is still four times wealthier than an average person living in China. It has a stellar education system. Three out of the top ten universities in the world and seven out of the top ten in Europe are here. And it's still a magnet for top talent. The UK ranks second only to the US as the most popular destination for international students. In per capita terms it attracts three times more international students than the US. So why does it struggle to translate all of that raw material into new technologies and big companies? The usual answers are mindset and culture.

The Dutch economist Arjo Klamer's schematic of caravan societies and citadel societies has become the standard explanation to understand the contrast in economic dynamism between the US and Europe. The US was founded by pioneers and immigrants willing to take risks in pursuit of new opportunities. This forged a 'caravan society' that is always on the move: mobile, dynamic and open to change. Europeans have been more rooted in their native lands: a static, 'citadel society' of settled, conservative people more concerned with preserving the old than exploring the new. These archetypes, deeply ingrained in the psychology of their respective cultures, shape economic behaviour and attitudes towards change and risk on the two sides of the Atlantic.

Hussein Kanji, an American venture investor based in London, says these differences in attitudes towards risk can be seen in how employees of US and European startups view compensation. These contrasting preferences have a downstream effect on repeat cycles of innovation that we see in the two regions.

'In America the percentage of a company that employees collectively own is much higher than in Europe. The typical seed round in the US would be about 20 per cent for employees. A typical European round would be 5 to 10 per cent max. In Europe employees undervalue stock and they want higher salaries whereas in America employees highly, highly value stock. And what ends up happening is, as companies do really well it's not just the executive group in those companies that gets rich, it's also the rank and file.

'So if you look at PayPal and you look at Skype, they were both roughly the same size when they first exited. PayPal sold for about a billion and a half dollars and Skype sold for $3 billion but it happened a little bit later, roughly the same size if you adjust for inflation. You look at PayPal, it created somewhere between 100 to 160 millionaires. And you look at Skype and it created eleven. And Skype was the bigger acquisition and PayPal the smaller acquisition.

'If you're an engineer and you're flush with cash, you don't tend to allocate capital to the typical diversified portfolio that your money manager might tell you. You end up using some of that money to become an angel investor and reinvest it into tech. This wealth is then recycled back into the system. But when wealth is concentrated in the hands of a few, it doesn't get recycled as much because there's less of it to use. And I think that's largely what's happened in Europe.

'Skype should have produced company after company, investor after investor, but almost all the proceeds went to eleven people and really not even a level eleven, just a handful of people. And that wealth got very concentrated, and you didn't end up with tons and tons of other companies and it didn't become nearly as virtuous as it should have.'

3

The idea behind hubs is simple: assemble a dense hive of diverse talents and then let serendipity and compounding work their magic. California has the Bay Area, Boston has Kendall Square, Seoul has the Gangnam District. London has that for finance, with the history of the Square Mile stretching back to Roman times, but not quite for tech. There used to be a small cluster of tech companies around Old Street, which then got renamed Silicon Roundabout, and then Tech City. The area was aggressively marketed by successive governments as their very own Silicon Valley, but in truth even its most ardent advocates sort of knew that it was weak sauce compared to what was going on in the big leagues. And then came King's Cross.

Two decades ago, King's Cross was a raw and brawny industrial badland of 'trains and more trains', strewn with rotting hulks of underused and derelict warehouses and railway lines. *Architecture Today* described it as 'an indigestible blockage in the fabric of London since the 1830s', an urban no-go zone that commuters took long detours to avoid. 'King's Cross was a total wasteland beforehand, surrounded by a lot of poverty on the estates and that was the history of King's Cross,' says Theo Blackwell, the Chief Digital Officer of London. 'It had a huge, huge drugs problem, huge drugs problem.'

Since 2006, King's Cross has been the site for one of the largest urban renewal schemes in Europe since World War II. It has taken fifteen years, £3 billion, and thirty-five architects to transform 67 acres of scorched warehouses and gritty railway land into a polished exposition of steel, glass and exposed brick. Heavy industrial structures still define the aesthetic but their shells have been repurposed for loftier pursuits. Gasholders, massive drums that used to store natural gas, are now cylindrical apartment complexes in skeletal enclosures of cast-iron frames with penthouses that go for high seven figures. The Granary Building, built in 1852, where once wheat was hauled in to be stored for London's bakers on horse-drawn carriages, is now reborn as Central Saint Martins, a world-famous art school.

Coal Drops Yard, built in 1850, has borne witness to some of the more salacious chapters of the district's history. 'The Pleasuredome' was for twenty-five years an outpost of two prodigiously seedy nightclubs, establishments that were reportedly so grimy that 'if you rubbed a wall, or if you squeezed past someone, your clothes would be black' from the soot that smeared everything; where DJs Skibadee, Nicky Blackmarket, and Judge Jules cut their teeth plying their trade in illicit, heaving, marathon super-raves with 'sweat literally dripping off the ceiling'. These haunts, where the city's hard-core jungle scene first took off in the nineties and whose vibe has been described in turns as 'edgy', 'dodgy', 'risky' and 'banging', have now stumbled into respectable midlife as £100-million shopping complexes where upmarket retailers peddle £100 candles, their place in the city's architectural firmament and polite society sanctified with two awards from the Royal Institute of British Architects.

King's Cross, or simply KX to those who know these things, erstwhile bastion of wide-ranging sketchiness and general delinquency, is now known, along with neighbouring Euston and Bloomsbury, as the Knowledge Quarter.

Here lies the superdense core of the city's thinking economy: a triple vertex of high culture, high commerce and high tech. Packed in a one-mile radius is an impressive anthology of knowledge-intensive institutions, among the highest concentrations anywhere in the world. Tenants include the 270-year-old British Museum which, with 8 million objects, is the largest museum in the world; the Wellcome Trust which, with a £37 billion endowment, is the third largest foundation in the world; and Google's billion pound campus, its largest outpost outside the US. The Crick Institute, a '£700 million cathedral to biomedical science', is also here. As are

Facebook, DeepMind, Samsung and hundreds of other public and private outfits.

Theo Blackwell, who has also served as an elected representative on Camden Council, the borough where King's Cross is situated, tells me that the redevelopment was primarily about culture, they just wanted the place to be less of an eyesore, and the tech sort of just happened. Even now they are mindful of unwittingly reproducing the social failings that plague some of the other high-profile tech districts.

'How do you ensure that the kind of benefits of the innovation culture that was being located there was also shared by the area around it?' he reflects. 'And the area around it is historically a working-class area: Somers Town, it's history goes back to Dickens, that's where Oliver Twist came from, so it was a question of assimilation and equity, so you don't create what we've seen in Palo Alto, which is a highly innovative, highly profitable tech community, but also not linked so well with the areas of poverty around it.'

4

King's Cross is a tech hub that's been superimposed on a transport hub. It is the busiest node in the London Underground and over 70 million passengers pass through its halls every year. Neighbouring St Pancras is an international gateway that connects the UK to the rest of Europe, sort of like a Heathrow of trains. It's the terminus of the Eurostar which runs direct high-speed rail services to Paris (~2 hours), Brussels (~2 hours), and Amsterdam (~4 hours).

Saul Klein, a well-known venture investor based in London, thinks that these rail links weave together a pan-European super region: the continent's answer to the Bay Area and the Pearl River Delta. 'We define the core area that we focus on as a four-hour train ride from King's Cross. And it's a geofence, it's not like a city or a country, it's a geofence defined by trains and transports. And we put a name on the geofence, New Palo Alto, and it turns out the New Palo Alto, after the Bay Area and Beijing has produced the third most unicorns in the world: 41 million people live in New Palo Alto. There's $110 billion of corporate IT spend. There's seven of the world's top thirty universities. And you've got all of the raw material for a world-class global innovation ecosystem.'

These train links make moving back and forth across international borders almost as easy as domestic travel. Saul can take a train out from London in the morning, hop over to Paris or Brussels to check out what's going on at his portfolio companies, and be back at his office near King's Cross just in time for lunch. This infrastructure backbone solidifies London's position as the de facto tech capital of Europe. The city is often the first port of call for tech companies from the continent looking to expand abroad as well as foreign companies trying to enter Europe. Many of the biggest names in tech that started elsewhere on the continent – Skype, King, Atomico and Index – can now be found in London.

These transport links also help connect the dots on a tech scene that is a lot more spread out than other places. 'Most of those high returning businesses historically in the venture capital industry in the US have come from the Bay Area; if you do the math in Europe, at least for us it's quite random,' says Sonali De Rycker. Research by Atomico, a venture fund, shows that the over 350 tech companies worth more than $1 billion that have been founded in Europe were created in more than 120 different cities across twenty-nine countries. Rail links pull this scatter into a coherent network facing in the direction of UK's capital. 'We're literally in the bullseye of the New Palo Alto,' says Saul Klein. 'We've got density that Palo Alto couldn't dream of.'

Works in theory. But how does it pan out in practice? I wanted to find out what would make an entrepreneur choose the New Palo Alto over the old one. And that led me to Marc Warner, the founder of Faculty, an AI company based in London.

Marc has advance degrees in Quantum Computing from UCL and Harvard and launched his startup in the very buzzy field of AI. He could have founded his company anywhere but thought London would be the best place to do it. I asked him why. 'In some sense Faculty has chosen to do what it does slightly because we wanted to be in London rather than the other way around,' he says. 'So if I wanted to be in London I think we had to take a view that we need to do things that are going to be hard to do in Silicon Valley.'

One advantage London has over the Valley is that startups don't have to compete as much with big companies for top talent. The UK has a very strong talent base in AI thanks to a high number of graduates coming out of schools like Imperial, UCL, Cambridge and Oxford and the ecosystem around companies like DeepMind. UK startups can tap into top-tier talent in a way that Valley-based startups just can't. Out there the competition is too strong. 'The talent war for AI is the craziest talent war I've ever seen!' Elon Musk tweeted in April 2024. Even the biggest companies are finding it hard to hold on to their best people. About half the AI experts OpenAI hired since its founding have since left the company to work for rivals, including its own co-founders. Big tech firms are poaching entire AI teams from their competitors. 'London offered a slightly more reasonable balance,' says Marc. 'It is obviously still a very competitive ecosystem but not as absolutely white hot as Silicon Valley. So we could get much more talented people in ways that a startup in Silicon Valley probably never could because of the amount of competition with big tech companies.'

Faculty uses machine learning to help people make better decisions, also known as decision intelligence. That sounds vague, but the company is solving a very tangible problem. The amount of data in the world is exploding. Ninety per cent of the world's data was created in the last two years. Every two years, the volume of data doubles in size. This poses challenges. All this data is worthless unless better tools are devised to do something useful with it.

That's where Faculty comes in. It uses AI to turn large amounts of data into intelligible choices.

If you've ever used Google maps you've benefited from decision intelligence. Whenever the app suggests alternative routes to help you save time, that's decision intelligence in action. The app is crunching millions of data points in real time to present the user with a simple set of easily actionable options. Marc says that those sorts of algorithms will soon be applied to a much wider range of activities. 'We think that businesses are operating in a pre-Google maps era, they have no map of the possible options. They're genuinely lost. Decision intelligence is going to give them Google maps. We're just going to see this kind of transition from people who were lost and now are found, as messianic as that sounds.'

Another way in which London compares favourably to the Valley is that it places startups near a lot of other players. It's a densely packed urban environment. The Bay Area has a high concentration of tech companies, but few other industries are based there and the government is all the way on the other side of the continent. London has all of that in one place. Colocation is efficient, reduces friction and puts companies like Marc's in close proximity to a lot of potential customers. 'You can go twenty minutes that way and be at the centre of government, twenty minutes that way and be at the centre of finance, twenty minutes that way and be at the centre of culture in a way that is actually relatively hard in Silicon Valley,' says Marc. 'Those twenty minutes become four or five-hour flights and sort of eight-hour round trips.'

London is not just dense, it's dense with a very diverse set of talents and disciplines. Alex Klein, British-American founder of Kano, a company that makes educational computing products, thinks this makes the city a more interesting counterpoint to the echo chamber of the Valley. 'A lot of their technology minds are doing Software as a Service applications to help businesses process their user data or they are joining Facebook and Google to help advertisers more effectively manipulate you,' he tells me. 'In London there is definitely plenty of that. But we are slightly different. It's like you definitely have people who are keen to build bike lights with lasers on them so people could see in front of them a little bit further. People who are building organic gardens. People who are ex-Dyson, who are great industrial designers, or ex-Nokia from Finland who are great firmware engineers, or people from Rockstar who are great video game designers. So I think the community in London is more multidisciplinary.'

The city also compares favourably to the Valley in other respects. Entrepreneurs feel at liberty to try new things. In America, the approach to launching and scaling companies has become standardized. There's a set way of doing things and even if an entrepreneur is not inclined to do them that way, investors expect that that's how they will be done. In newer ecosystems like London where processes are less set in stone, there's a lot more room to play around and experiment. 'In Silicon Valley there's a particular set of beliefs around hyper scaling, blitz scaling, all these kinds of ideas where it's very, very focused on a particular playbook and it is quite disconnected from the wider world,' Marc continues. 'There are upsides and downsides of London and other places outside Silicon Valley, but it does let you have a bit more freedom. And I think we've come to understand that if it turns out that decision intelligence is as important as we think it's going to be, we would never have been able to find that out in Silicon Valley. I think that much is true.'

5

There's a flip side. The attribute that consistently comes up as London's greatest strength – all good things all in one place – is also often held up as the wider country's biggest failing. All that concentration begets a strong Matthew effect – 'for whosoever hath, to him shall be given, and he shall have more abundance' – that is, London's advantages compound to the detriment of outlying regions. London has more than twice as many billion-dollar companies and gets almost twice as many venture investments as the rest of the UK combined. The distribution gets even more skewed when you add Cambridge and Oxford to the mix. These three cities, known as the golden triangle, get 80 per cent of all investments and have 80 per cent of the billion-dollar companies in all of Britain.

'And that's the negative on London,' says Eben Upton, the inventor of the Raspberry Pi. 'Having a load of tech in London, which is a cab ride from Westminster, does make it a little bit easy for the politicians to forget. It's quite nice in America where the centre of political power and the centre of technology power are on opposite sides of the country. In London, it's a cab ride apart. And then people forget that there is stuff elsewhere very quickly and then they under-invest and then you let genuine opportunities for regional policy go by the wayside.'

The problem of regional disparities is much older and much wider than just the tech sector. From life expectancy to employment to education, London fares significantly better than the rest of the UK, particularly the North East and Wales. 'Why Hasn't UK Regional Policy Worked?', a report co-authored by Ed Balls, the former Labour shadow chancellor, notes: 'The UK has some of the highest regional inequalities of any advanced country. Today, these are larger than those between east and west Germany and north and south Italy. New technologies, global competition, the loss of old industries – and the failure to support new ones – have all driven that divide.'

London contributes a quarter of the UK's entire GDP, a much higher proportion compared to just about every other metropolitan area in any other big country in the industrialized world. Compare that to New York, Shanghai and Berlin, the largest urban economies in their respective countries, which are responsible for only 8 per cent, 3.7 per cent, and 4.7 per cent of their national GDPs. If London were an independent country, its economy would be among the twenty largest in the world, ahead of places like Switzerland, Sweden and Turkey. But such is the extent of UK's economic monopolarity that without London the rest of Britain would in per capita terms be poorer than every single state in the US.

Resentment at this tale of two Britains – one that is urban, cosmopolitan and upwardly mobile, and the other rural, inward-looking and economically stagnant – was a major force behind Brexit with the Leave campaign seen as appealing to the neglected north, while Remain was portrayed as representing elite southern interests. When Boris Johnson was elected prime minister in 2019, he was elected on two major policy planks: 'Get Brexit Done' and 'Levelling Up'.

'I've got nothing against London, but it feels like we need a regional policy,' Eben continued. 'Britain already has a lot of trouble because it's in a situation where London and Cambridge are the only two bits of the UK which are contributors to the exchequer, everywhere else is a net draw on the exchequer, and that's not desirable. And if Cambridge just becomes a commuter suburb of King's Cross, then that would be a shame because if you can't even sustain regionalism at a distance of forty-eight minutes then London just becomes a city state. I mean London becomes Singapore. It's not Britain any more, is it? It's just a city state that currently hasn't figured out how to decouple itself from its surroundings and that's not good.'

These regional disparities stretch back to the 1800s. During the interwar period, still the heyday of empire, when Britain held sway over a quarter of the world's population and a quarter of the earth's entire landmass, living standards in many localities at home were not much different from those in far off colonies. In 1937, the UK government appointed the Royal Commission on the Distribution of Industrial Population to investigate the causes, consequences and remedies for economic concentration. The Barlow Commission Report, published in 1940, concluded: 'The contribution in one area of such a large proportion of the national population as is contained in Greater London, and the attraction to the Metropolis of the best industrial, financial, commercial and general ability, represents a serious drain on the rest of the country.'

More recently, in 2022, the UK government issued 'Levelling Up the United Kingdom', an influential white paper that provided the most comprehensive articulation yet of post-Brexit Britain's signature policy agenda, which included a proposal to transform 'derelict urban sites into beautiful communities' by replicating King's Cross-style regeneration projects in twenty other locations.

But even as successive governments have in their own way declared levelling up a major policy priority for well over a century, the regions have progressively only ever levelled down. Since the 1980s, industrial towns in the north have been hollowing out as manufacturing jobs move abroad while the simultaneous boom in financial services has buoyed London's fortunes. And now the tech boom is amplifying this divide. London is no longer just a major financial centre. It is also a formidable tech hub: it has the fourth largest cohort of billion-dollar tech companies anywhere in the world, after San Francisco, New York and Beijing.

In the US and China the centre of gravity of the tech economy is spreading from the traditional powerbrokers to more regions: Austin and Miami, Hangzhou and Guangzhou. In the UK the trend points to increasing consolidation. 'The UK used to have a lot of tech clusters,' says Eben. 'It had Silicon Fen, Cambridge, it had Silicon Glen, it had a big electronics manufacturing cluster up in Dundee, it had a lot of semiconductor stuff over in the Bristol area,' but, he says, over time their relevance has waned as London has pulled more and more functions into its orbit.

6

Just about all the tech companies that get any amount of attention outside the UK – DeepMind, Revolut, Arm – are based in the golden triangle. I was curious to hear what it's like to run an ambitious startup outside the mainstream and Saul Klein pointed me in the direction of Radix, a startup based in Stoke-on-Trent, a small town of only about 250,000 residents that, even if it's only a short two-hour train ride north of London, feels like a world apart from Britain's capital.

There are many ways to describe Stoke. It has been called a symbol of left-behind Britain. The city is practically a textbook case of a former industrial boomtown, the world centre for ceramics manufacturing in the eighteenth and nineteenth century, that has got the raw end of globalization. It's also sometimes called the Capital of Brexit: a full 69 per cent of its voters voted to leave the EU, more than any other large UK town or city. And at least one person on Reddit called it 'a top place to smash a glass bottle for no reason'. In other words, exactly the sort of dystopian outpost from which a brilliant but reclusive engineer would plot the demise of bitcoin, now worth over a trillion dollars, and blockchain, the technology on which it is built.

Dan Hughes was born and raised in Stoke where, from the age of four, he picked apart machines and put them back together again to figure out how they worked. 'My true skill is being able to look at a problem and be able to find interesting non-obvious ways to solve that problem,' he says. He skipped college and went straight into the gaming industry, then started his own company where he worked on software for NFC, the technology now widely used in mobile payments. Flush with cash after selling his startup, he decided to jump into bitcoin in 2011, when crypto was still in its infancy, and worked on fixing what he thought was its Achilles heel: scalability.

Dan, who is a Y Combinator alum, thinks that bitcoin is a revolutionary idea but the infrastructure on which it's built just isn't designed to handle billions of transactions. 'After a little while it felt like, okay, bitcoin is a horse and I'm going to try and stick a jet engine inside it and make it fly,' he says. 'But it's going to be bad for the horse and it's probably not going to work very well. And so that was when it dawned upon me that okay, what bitcoin really is, is the first step and it's kind of shining a light on the path to take. But it's a Model T Ford. And today we drive around in Teslas. So then it was obvious that, okay, learn from bitcoin, learn what Satoshi made there, how he solved some of the problems, and then open a new file and start with a new architecture.'

Radix is an alternative to blockchain. Both are Distributed Ledger Technologies (DLTs). Distributed ledgers function similarly to databases in storing data, but unlike traditional databases, which are centralized, DLTs distribute data across many nodes. This distributed nature makes tampering with the data much more difficult because there is no single point of attack. Any attempt to alter data on one node can be detected and corrected by comparing it with copies on other nodes.

Blockchain stores transactions on blocks (the 'block' part), which are then processed sequentially (the 'chain' part), leading to bottlenecks and slow transaction times as the network expands. PayPal handles around 200 transactions per second (TPS), Visa processes 2,000 TPS and has a capacity to go up to 56,000 TPS. Bitcoin can only manage about 7 TPS. Ethereum, the world's second most popular cryptocurrency, only clocks in at around 15–20 TPS.

Dan says this makes the blockchain an unlikely candidate for mass adoption. 'Bitcoin, from a technical point of view, if not already obsolete, is going to be obsolete real soon, even if it's not made obsolete by what we're building, it's just such a fast-moving space.' Radix uses a different approach which allows multiple transactions to be processed in parallel rather than sequentially which significantly improves scalability and transaction throughput. It does everything that the blockchain does, but faster, and even has its own currency, called XRD.

DLTs are not just about financial transactions or even smart contracts. The plumbing of the entire internet will eventually be re-architected around crypto networks. In the same way that crypto has created programmable money outside government control, it will, in the idealism of the Web3 movement, also be deployed to wrest control away from other forms of central authority: finance without banks, news without newspapers, social media platforms without social media companies.

'These big platforms, so your Airbnbs and your Ubers and social media websites and all that stuff, those companies in the extreme case won't exist in fifteen years,' says Dan. 'And they will own nothing either, because you will have this very tight integration of the Web3 social network, Web3 file storage, Web3 Ubers, where everything is very peer to peer. And the citizens of that economy own their data, the rights to the data, their interactions, and can monetize them or whatever, however they want.'

Good stuff. So, why Stoke? 'It's comfortable in Stoke, it's quiet,' says Dan. 'I've spent some time in London, it's impossible to focus because there's just always noise constantly.' A lot of it is just down to personality traits. Dan is rare among founders in that he willingly gave up being CEO of his own company to focus on research and development. 'It doesn't really matter that I'm here in terms of the stuff that I'm doing,' he says. 'So it just works. It just works for me basically.'

7

Another big breakthrough to come outside of the mainstream in the UK is graphene. In 2004, two researchers at the University of Manchester, Andre Geim and Konstantin Novoselov, isolated a 2D material composed of a single layer of carbon atoms arranged in a hexagonal lattice. It is the thinnest material that exists, only one atom thick, and also the strongest. It is about two hundred times stronger than steel and a thousand times lighter than paper. A single sheet of graphene that covers an entire football field would be practically invisible, weigh less than a paperclip, and yet be strong enough to support the weight of a car.

Graphene is in all respects a supermaterial. It closely resembles graphite, the stuff inside pencils, and is made up of carbon, the fourth most abundant element in the universe, making it a potentially eco-friendly alternative to plastics and metals. It is not just the strongest and lightest material that exists, it is also the best at conducting heat and electricity, about a thousand times better than copper. These versatile properties make this 'material of the future' a good fit for a range of applications.

Graphene is a better conductor than silicon and can replace it in semiconductors which would help along the transition from microelectronics to nanoelectronics which exist on a molecular level. That will open up the door to a world of flexible, printed devices; imagine a television or a tablet that can be folded like a piece of paper. It can also be used in batteries. Graphene-based batteries would be lighter and conduct electricity faster than the lithium-ion ones in vogue today, which means that everything from phones to electric cars can go for weeks between charges.

Graphene is not some hypothetical material that exists only in science fiction, it's already in use today. In retail, graphene-based circuits are printed on security labels attached to items in shops which set off an alarm if the item leaves the premises without authorization. Ford makes plastics for its vehicles that are 0.5 per cent graphene, increasing their strength by 20 per cent.

Graphene is difficult and expensive to manufacture on an industrial scale and that has held back its mass adoption. But the progress of technologies from discovery to deployment is often slow. MRI was developed in 1973, three decades after scientists first figured out the mechanics behind how it works. Silicon was purified in 1824, more than a century before it gave birth to the semiconductor industry.

Andre Geim and Konstantin Novoselov were awarded the Nobel Prize in Physics in 2010 for their work on graphene. The fact that such a major discovery can come from outside the golden triangle shows the UK's bench strength in research institutions that can produce top-tier research runs deep. The US has no peer at the very top end of higher education. But in an observation made to me by a number of researchers, on average the UK's research universities fare better than those in the US. 'It's an interesting philosophical question,' says Francesco Sciortino, the founder of Proxima, a nuclear fusion startup based in Munich, who is a graduate of MIT in the US, Imperial in the UK, and EPFL in Switzerland. 'Given some resources, is it better to concentrate them into a few extremely good people or is it better to take many more shots on target? The latter is what Europe does more with a few exceptions here and there.'

The UK has plans to improve its standing in tech by bolstering the substrate of scientific capabilities on which it is built. In 2023, the government unveiled the Science and Technology Framework to turn the country into a scientific superpower by 2030 and created a new Department of Science, Innovation and Technology to lead this effort. An interesting wedge has opened between the US and the UK on the question of the extent to which the scientific approach should guide the development of future technologies. In the US, the frontier for this sort of big research has unmistakably moved to startups and big tech companies. In the UK, there's still a lot of scepticism at the idea that commercial motives should play such a disproportionate role in creating powerful technologies like AI. 'It's too important a technology to only be the preserve of the tech giants,' David Barber, the Director of the UCL Centre for

Artificial Intelligence, told me. 'You do need independent bodies such as academia to keep check on this stuff.' DeepMind's founder, Demis Hassabis, has often drawn a sharp contrast between his preference for a scientific approach to building AGI as opposed to the more freewheeling hacker ethic prevalent in Silicon Valley.

The country has also unveiled more unorthodox plans to complement these traditional avenues to realize its goal of becoming a scientific superpower. 'If you want wildly different outcomes from the ones you're getting, you're going to have to try some wildly different processes,' says Matthew Clifford. Matthew was only 36 years old when he was handed the very heavy responsibility of leading the UK's Advanced Research and Invention Agency, or ARIA, in 2022. The newly established organization, loosely modelled on DARPA, has an £800-million budget to fund radical moonshots that can create internet-type advances. The aim is to get behind technologies that are too early to raise money from the market. 'Venture capitalists are good at taking market risk but not good at taking technological risk,' Clifford tells me. 'We only fund things that might not be possible.'

In addition to providing direct financial support, ARIA is also trying to change the rules of the game by rejigging professional incentives for young researchers so they can take more risks earlier in their working lives. 'Early career scientists are very strongly incentivized to pursue incremental work that they can more or less guarantee a publication will come out of,' says Clifford. 'As an early career scientist, it's kind of insane to try and do something ambitious because if it doesn't work, you don't get any publications, you'll never get funded. And so, rather than have funding follow predictable track records of incremental progress, we are trying to shortcut that and let the most ambitious people work on their most ambitious ideas now and take the kinds of risks that pursuing conventional funding will not reward.'

8

That's the past and present of UK tech. But what should the future look like? Preferences are divided. Some think that the UK tech sector needs to be more like the one in the US and that means scale. 'What's my yardstick of success? I'd like to see a British Alphabet, I'd like to see a British Microsoft,' Jeremy Hunt, then Chancellor of the Exchequer, told the *Financial Times*. 'I'd like to see a homegrown company with a trillion-dollar cap, with a big global position.'

That just about sums up the mainstream sentiment in tech circles in the UK. But there are pockets of dissent. They ask a question that is the closest thing to blasphemy in the tech industry: is a trillion-dollar company a desirable policy outcome?

The question is especially relevant at a time when even the US is grappling with whether its own tech giants are a symbol of its economic strength or a symptom of antitrust failure. Is a trillion an excessive confirmation of capitalism's extreme competence at creating abundant value? Or is it a numerical summary of the extent to which regulatory responsibility has capitulated to corporate power?

The ten largest American tech companies are now collectively worth a third of the entire US stock market. The combined value of just these ten companies exceeds the entire GDP of every country in the world except the US itself. The question is no longer whether too much economic power is concentrated in the hands of a small number of players. It *is* concentrated in the hands of a small number of players. The question is what to do about it. And this question is no longer being asked just from the activist left but from decidedly right-leaning voices from within the mainstream of the US tech community.

'I think that these big tech companies like Google and Microsoft do need a check on their power,' David Sacks noted in the All-In

Pod in the summer of 2023. 'Someone does need to cut these big tech companies down to size. They are giant monopolies and they do need to be restrained and controlled or they will basically consolidate the whole tech ecosystem and abuse their market power.'

At a time when even the US is second-guessing the social utility of trillion-dollar companies, should the UK make it an explicit policy priority to manufacture its own? Is the presence of hyperscale companies even a useful metric to measure the innovative capacity of an ecosystem? Is the health of the economy better served by a few very large companies or many small and medium-sized ones? Does size even matter?

'I don't know if it's a problem. I don't know if it's a problem. You can clearly have a very successful economy without growing a large number of companies to billion-dollar scale,' says Eben Upton.

Over a long conversation, Eben shares how the challenges of doing a lot with very little are often more interesting than the challenges of doing a lot with a lot and how it's not necessary to achieve superscale proportions to produce superscale outcomes.

'Look at Israel, tiny country, a very strong ecosystem based around growing medium-scale companies and selling them to American companies. And that's a fine way to run a country. And if we ran our country, if our tech industry was like that, that would be fine.'

But that model – grow medium-sized companies and sell them to bigger companies abroad – comes with its own baggage. It might work for smaller countries. But in the UK the idea that the purpose of the most advanced sections of the country's economy is to cook up companies only to serve them to the Americans, feeds into a narrative of decline and stokes broader anxieties about the country's diminishing place in the world. Pocket superpower, 51st state, American sidekick and all of that. In influential sections of the commentariat, for whom the glory days of Britain have not yet faded from memory, the whole point of investing in a tech sector is for it to help the country compete with other major powers and not merely to confirm its position as their adjunct.

This sentiment found its fullest expression in 2020. The Google/DeepMind acquisition in 2014 met little resistance and was largely seen as a welcome validation of the UK tech scene. But just six short years later the mood had completely changed. In a reminder that timing and market conditions are everything, Nvidia's proposed takeover of Arm in 2020, almost an exact parallel of the Google/DeepMind deal made six years earlier, only bigger, with an American suitor shelling out $40 billion, the largest deal in semiconductor history, more than eighty times what Google had paid for DeepMind, was met with outright hostility across large cross-sections of expert opinion in Britain.

A lot of that had to do with what Arm does. The company is practically the definition of a strategically significant, globally relevant tech company. Arm was founded in Cambridge, England in 1990 in a partnership with Apple. Arm designs chips and then licenses those designs to other companies which manufacture the actual chips. If chips are like buildings, then Arm is like the architect of those buildings.

Arm, which has sometimes been called 'the world's best kept technology secret', is arguably the most consequential tech company ever to come out of the UK. Its presence is ubiquitous but invisible. Over 95 per cent of all mobile devices in the world use processors designed with Arm technology. It is a critical node in the semiconductor supply chain and one of the world's most valuable companies based entirely on intellectual property.

When SoftBank acquired Arm for $34 billion in 2016 – at the time the Japanese conglomerate's largest acquisition ever – the company's billionaire founder Masayoshi Son said that he expected Arm to be more valuable than Google. And that's why the Arm deal rankled with popular sentiment in the UK. The novelty of American companies taking interest in UK firms had long worn off and unlike the DeepMind acquisition this deal was seen more as the UK losing a national asset rather than gaining a wealthy foreign patron.

In an article in the *Financial Times* titled 'Arm's destiny vital for Britain's future', John Thornhill, the paper's technology editor, wrote that this was no ordinary commercial transaction and the entire country's technological independence was pegged to Arm's fate: 'if national sovereignty in the twentieth century was magnified by military hardware – tanks, battleships and nuclear missiles – it is increasingly empowered today by civilian software – intellectual property, data and computer code.' He urged the UK government to step in to block the deal arguing that 'Britain's claims to be a sovereign nation in a digital world will further evaporate if it fails.'

Hermann Hauser, one of Arm's own co-founders, was even more strident in his disapproval. In a letter to the *Financial Times* he wrote: 'Now we are about to witness one of our last great European technology companies with a dominant position in mobile phones becoming part of the US trade armoury. Whether we are allowed to use our own British-designed microprocessors in the UK and Europe will be decided in the White House rather than in Downing Street. This is a major step towards the UK becoming an American vassal state. It must be stopped.'

9

I reached out to Arm's CEO Simon Segars in the fall of 2020 just as this drama was unfolding to find out how he felt about this very public backlash against what for him was the most important decision he made as an executive and one that would certainly define his legacy as CEO.

Simon, an Englishman born in Basildon, a small town east of London with a population of around a hundred thousand people, is every bit as understated as the company he ran for almost a decade. He joined Arm as its sixteenth employee in 1991 when the startup was only a few months old. In the subsequent three decades he saw the company go from a barn in Cambridgeshire to becoming the most pervasive computing platform in the world, elevating his own fortunes from an entry level engineer in an obscure startup to the CEO of a global business.

Simon was relaxed and even reflective as he shared stories from the formative years of Arm, with gentle humour about the competitive advantage bestowed on them by their British accents which helped the unproven startup close deals as far away as Japan. But he couldn't hide his frustration when the topic turned to the controversy around the Nvidia deal. He thought his critics were naïve to the realities of modern capitalism in which the question of who owns a company is a lot more complicated than they would make it out to be.

'I've got to say I struggle with this technological sovereignty thing,' he said. 'People say, oh it's a real shame that Arm got bought by a Japanese company and I point out to them well before we were acquired, more than half our shares were owned by non-UK institutional investors. So were we a British company before?' He added: 'The fact is that when the company was set up there were three joint venture partners, one of whom was British and the other two were American. So actually from day one we were minority owned as a British company.'

In a telling sign, Simon spoke to me from San Francisco, the first Arm CEO to be based outside the UK. It was a fact I had heard others mention in mildly disapproving tones in my conversations back in Britain. But Simon shrugged it off. It's natural for companies to rethink their geographic emphasis as they go through different stages of growth. Plenty had done it: Skype, UiPath, Miro, it's a long list.

'I've always described Arm as a global company that happens to be born in Britain,' Simon said. He thought that the criticisms directed at him had less to do with sound economic logic and influenced more by the broader political currents sweeping the country. 'I think the last sort of five, six years in the run up to Brexit and post-Brexit has made UK a lot more inward-looking and less kind of global in its outlook,' he said, lamenting that the country of his origin just didn't have a healthy culture of celebrating success.

Later that year regulators on both sides of the Atlantic raised concerns about the Arm acquisition. In the US, the FTC blocked it on antitrust grounds. In the UK, regulators invoked more dire national security concerns. The deal unravelled. Simon, who had bet his entire leadership on the acquisition, departed Arm soon after.

So what did the regulatory action achieve? If in the US the objective was to curtail Nvidia's market power then blocking that one acquisition, relatively minor in the Nvidia scheme of things, did close to nothing to break the company's momentum and it went on to cross the trillion-dollar mark three times over in the next three years.

And it's an open question if the UK managed in any meaningful way to 'keep Arm British'. The company, which was Japanese-owned to begin with, skipped a London listing and went public in New York in 2024. The deal was scrapped to stop Nvidia from becoming a monopoly and to prevent Arm from leaving the UK. Then Nvidia became a monopoly. And Arm left the UK. So, what was the whole fracas all about then?

I reached out to Hermann Hauser when Arm went public in NYC to find out how he assessed the failed acquisition in hindsight. He stuck to his guns. 'I'm delighted that the Arm takeover by Nvidia did not happen,' he said. 'It would've created yet another American monopoly that we could do well without.' He saw the Nasdaq listing as the less bad alternative. Europe's tech sovereignty is better preserved if Arm is publicly listed with distributed ownership, even if it's in the US, than if such an important company was allowed to be controlled outright by one American tech giant. 'It remains a UK headquartered company, in that sense it's still a European company.' The scuttled deal may have been a boon for Arm's investors. Nvidia had offered $40 billion for Arm. It is now worth five times that in the public markets.

But the central policy question remains unresolved: the next time a big American company swoops in to scoop up a prized British tech asset, as it inevitably will, should the role of regulators be to bless that union, like it did with DeepMind, or to prevent it, like it did with Arm? Or, to go back to the original framing, is the larger ambition here for the UK to build its own trillion-dollar companies or to make peace with playing a supporting role to the ones in the US?

The trend is unmistakably towards the latter. The UK doesn't have any new heavyweight contenders in tech or any other sector for that matter and that has been the case for a very long while. Britain has only three publicly traded companies that rank among the world's hundred largest. That's fewer than Switzerland, Germany, France, Canada and Japan, let alone the US and China. The three – BP, Shell, and HSBC – are neither new nor tech, with the history of all three old economy companies stretching back at least a hundred years. Angus Hanton, author of *Vassal State*, writes that 'it is notable that the FTSE 100 had more domestic tech firms in the year of its founding, 1984, than it has today.'

A lot of that has to do with the fact that high growth assets that can potentially reach that scale are, like DeepMind, quickly snapped up by foreign buyers. The battle over the Arm acquisition was still in full swing when in May 2021 Snap quietly bought Oxford-based AR startup WaveOptics. In subsequent years the foreign buying spree has continued unimpeded. Thoma Bravo bought cybersecurity firm Darktrace for $5 billion, Schneider bought industrial software maker Aveva for $11 billion, Etsy bought social commerce platform Depop for $1.6 billion.

Short of full-scale acquisitions, foreign players also hold substantial stakes in UK tech companies. According to Tech Nation, the recently dismantled and then resurrected body responsible for promoting the country's tech sector, one in three dollars offered by institutional investors to UK tech startups comes from the US. It notes: 'On the one hand, this could be seen as a sign of strength and burgeoning international reputation for investment returns in UK tech, but on the other, this may be seen as potentially problematic if UK tech firms with significant profit and influence are owned by non-UK actors.'

10

The question of how the UK can build its own tech giants isn't as simple as saying, well, we'll just not let others buy our companies or invest in them and then the UK will have its own Google. It's a bit more complex than that. The case of Graphcore is instructive.

Graphcore is a Bristol-based company that makes chips for AI applications. It's a pocket of the tech industry that's been getting a lot of attention lately with the US, EU and China each putting aside $50 billion in public funds to shore up their domestic chips industries.

There are a couple of reasons for this. The first is that chips, the brains of machines, the most critical components of all computing devices, are of all the things that have been called 'the new oil' probably the most deserving of the title. China spends more money on importing chips than it does on importing oil. No government wants to compromise their technological sovereignty by being overly reliant on foreign players for their uninterrupted supply of this critical resource.

The second is that the industry is going through something of a phase transition. Ever since chips were invented in the 1950s, improvements in their performance have largely been a function of the number of transistors packed inside them. Transistors are tiny switches that are the basic building blocks of modern electronics. The first commercially produced microchip, the Intel 4004, released in 1971, had 2,300 transistors etched into a chip that was about half an inch in size. Nvidia's H100 chip, released in 2023, has over a billion transistors packed into the same real estate. That's a difference in density of over 500,000x achieved in just fifty years.

But that impressive progress is now running into a major obstacle: it is becoming harder and harder to develop smaller and smaller transistors. The transistors in Nvidia's H100 chip are only 20 atoms wide. Soon miniaturization will run into the hard wall of physics. Researchers are developing alternative approaches to designing microprocessors whose performance will not rely solely on squeezing more transistors into chips.

This shift to new types of microprocessors doesn't happen very often. Hermann Hauser points out that it has only happened three times: the first when low-powered chips were developed for mobile phones, second when Graphical Processing Units (GPUs) were created for high-intensity video processing, and the third is now. This opens up the opportunity for new entrants to shake up the industry and make a play to be the next Intels and Nvidias of this space. Just the kind of big open white space the UK would need to build its trillion-dollar contender.

And that's where Graphcore comes in. The company makes chips called Intelligence Processing Units (IPUs) which compete directly with Nvidia's GPUs. Nigel Toon, Graphcore's CEO, tells me that their IPUs are designed from the ground up for artificial intelligence applications which require their own specialized hardware. 'The data structures that you are working with in AI are very different to conventional compute,' he says. 'It's incredibly complicated, highly parallel, mathematically what you would call high-dimensional information; so how you do processing on that is very different from the conventional rather linear approaches that you would typically use in computing.'

Nigel, who also serves as a director with UKRI, the UK's main public body for funding innovation, was optimistic that the next big shift in semiconductors could come out of the UK. 'There is no question that Europe and the UK are up there with the best in the world at making fundamental breakthrough research,' he said. 'I'm very bullish in terms of what Europe is going to achieve over the next twenty, thirty years versus the US and versus China actually.' He added: 'If you had the freedom to live anywhere, Europe's a really great place to live from a quality of life point of view and diversity and all of these things. So if you can build your business in Europe, why am I going to relocate to Silicon Valley? I'd rather do it here if I can.'

That conviction has been severely tested. Graphcore, like just about every semiconductor manufacturer out there, has struggled to mount an effective challenge to Nvidia. And in the company's own assessment that largely comes down to a lack of financial support.

In the spring of 2023, just a few months after Nigel made that case for why the future of tech belongs to Europe, the UK government announced the launch of Isambard AI, a billion-dollar mega project to build Britain's fastest supercomputer not far from Graphcore's headquarters in Bristol. But the project made no provisions to deploy Graphcore or any other UK-based company's technology in its compute infrastructure.

Nigel wrote an open letter to the prime minister which argued that unless the UK companies were brought into the project its financial benefits would accrue to US-based companies. 'Too often we have seen British-made innovation leading the world, only to be edged-out or bought-out by overseas rivals,' he wrote. He told the press that if the UK didn't prioritize its own tech companies then it risked 'heading towards a world of tech colonization' by US behemoths like Nvidia, adding that this could cause the UK to 'lose some of our tech sovereignty'.

This was just the beginning of Graphcore's troubles. Later that year it had to pull out of China, a major growth market. Nvidia is prohibited from selling its high-end chips in China, leaving the market wide open for Graphcore. But those export restrictions were soon levied on Graphcore as well, depriving it of a major source of revenue. Layoffs soon followed. As losses mounted the company filed accounts stating that it faced 'material uncertainty' that it could remain a going concern. Unable to raise more money from investors, Graphcore sold itself to Japanese giant Softbank in the summer of 2024 for $600 million, well below its peak value of $2.8 billion, and below even the $700 million that it had raised since its inception.

It's not unusual for tech companies to be in the red for a long time while they hone their technology and figure out a business model. Reddit did not make any money in its first twenty years. US-based tech companies can indulge in those sorts of practices because they just have a lot of money sloshing around. European companies operate in a different environment.

The Graphcore example illustrates the Catch-22 British companies face when it comes to raising money. In the absence of sufficient domestic sources of risk capital they have to turn to foreign investors. If they take the money, they sign-off their high potential trillion-dollar contenders to foreign buyers. If they don't take the money they might not survive. And the one alternative to private capital, the government, is not stepping up to the responsibility in ways that many think it should.

'The biggest problem is: "I got cool stuff, but the government's not buying it,"' says Nathan Benaich, a London-based venture capitalist who is also an investor in Graphcore.

Semiconductors are the deep end of risky, and private investors are usually hesitant to dive in. It can take years and tens of millions of dollars to figure out if the technology even works. This is where governments can make a difference. Benaich says that the US semiconductor industry took off because the government was willing to buy their wares at a time when the companies were too risky for private investors to invest in and their products were too expensive for other companies to buy. That laid the foundation for not just the US domestic chips industry but arguably all of Silicon Valley and the subsequent two generations of US supremacy in new technologies. But, as the example of Isambard's procurement shows, that practice of government acting as the first customer of unproven startups is less prevalent in Europe.

'The European knee jerk reaction to a market problem is to throw money at it,' says Benaich. 'You'll see this all the time, of funding gap here, funding gap there, then they start a fund and go invest in it. But I think the solution to these funding gaps is more like if there's an apparatus that would buy products from companies, then they would show traction and then investors would invest in them. It's pretty basic. People buy good stuff and the market will solve itself. So I sometimes wish instead of all these funds, we'd have advanced procurement agencies which would just rapidly acquire stuff.'

II

The underlying assumption to Europe's fixation with company size is that its tech industry needs to look more like the one in the US. But what if the continent's future lies not in being the same but in being different? What if the relevant distinction here is not the one between big companies and small companies but between good companies and bad companies?

In early 2010, Saul Klein, then a partner at Index, Europe's best-known venture fund, agreed to be shadowed by a journalist from the *Financial Times* for a week as he shuttled between Estonia and London to tend to his firm's investments. The picture that emerged was of a man on a mission to grow a Google-sized success story out of Europe. That Saul would agree to lay his work life bare like this to an outsider struck the journalist as a decidedly un-European act of openness and transparency, betraying habits of mind more common to the new world than the old continent.

And that to Saul was precisely the point. If Europe was serious about growing its own tech giants then the task at hand wasn't just about improving its competence in new technologies; the continent would have to submit to an entirely new way of thinking. 'Nobody would ask you why [in America], they would ask you how,' Saul noted. 'I want to bring that Silicon Valley mindset to Europe.'

There was a lot that qualified Index to be in the business of importing mindsets. The firm was founded in Switzerland in 1996 at a time when technology and Europe were still obverse concepts. In its early years when the firm's founder, Neil Rimer, went around raising funds from European investors he would start his presentation with the slide 'What is Venture Capital?' When he gave the same presentation to investors in the US, he would replace it with the slide 'What is Europe?'

And there lay the arbitrage that became the foundation for the firm's early strategy. Index had two headquarters: one in San Francisco and one in London. Saul made the pilgrimage to the Valley every quarter. Other partners went multiple times a month. They thought of themselves as something of a conduit for the Valley's way of thinking in Europe, with the balance of that liaison tilted heavily in one direction. 'We should forget about how to compete with the US,' Saul had told the *FT*. 'We should start thinking about how we can learn to play well with them.'

A lot had changed for Saul in the twelve years since he spoke those words and when I caught up with him in the winter of 2022. He had left Index to launch his own fund, LocalGlobe, a prolific investor in European startups. He picked up an OBE and sat on the UK Government's Council on Science and Technology. But the address on his business card and the titles next to his name were minor changes compared to the near total revision in his outlook towards the Valley's approach to building companies.

'For many years at Index we exported the Silicon Valley mindset and brought it to other regions,' he told me. 'And you look all around the world and that's a pretty successful model. And I guess the thing that I started to realize is that the Silicon Valley mindset doesn't always end well and there are aspects of that mindset and mentality that are actually dangerous and don't take people into account.'

The decade following the dot-com boom revealed the strengths of the Valley model. But subsequent years have also laid bare the costs of that approach. Surveillance capitalism and privacy concerns, algorithmic amplification of political differences, monopolistic tendencies and anti-competitive behaviour, outright fraud at companies like Theranos, toxic culture at companies like Uber. For Saul this pattern of bad behavior amounts to a systematic crisis of values that pervades the entire industry. 'It's like pick your company,' he says.

And thus Europe's big opportunity is not so much in replicating the Valley model as it is in going beyond it. The global tech race is usually framed as a two-way contest between the US and China with Europe as a distant also-ran. But that way of looking at things neglects the role that differentiated value systems are going to play in the next phase of technological development.

The past quarter century was marked by unchecked growth of big tech in the world's two largest economies. But the zeitgeist has now turned against them. As public backlash pushes back on big tech in the US, and state power takes on tech power in China, Europe can play the long game by building a tech ecosystem that is profitable but also long-term sustainable in that it is grounded in better value systems and more acceptable to the communities in which it operates. In other words: pigs get fed, hogs get slaughtered.

'When we talk about a New Palo Alto, the new bit is that bit,'

Saul continued. 'The difference is this, a values-driven ecosystem. And as a result, I think it is going to be at least as powerful in the next twenty to thirty years as the Bay Area or Beijing and possibly more because of the values. If you believe values are going to drive incremental value, then it is not about blitz scaling and go big or go home, or single stakeholder capitalism where the single stakeholder is the shareholder – the investor or the founder in Silicon Valley – and it's not single stakeholder capitalism in the way that it is in China – where the stakeholder is the state.'

It's not just the idealism of one man. I came across this sentiment often in Europe, that the continent is animated by more humanist values and cares more about rights, sustainability, equity and respect for privacy, and all of that can and should reflect on how it does business, and that is what will in the longer run set it apart from its competitors. These are not abstract debates. They have played out in very real ways in corporate boardrooms.

12

DeepMind has spent much of the past decade second-guessing its decision to sell itself to Google. The issue isn't the price tag. It's the loss of independence. It just doesn't trust that its powerful technology will be used responsibly by its corporate parent. In 2017, at a retreat at the Macdonald Aviemore Resort in Scotland, DeepMind's leadership revealed to its staff its plan to separate from Google. These efforts to gain more autonomy were informally referred to as 'Watermelon' by the company's employees and then later formally renamed 'Mario' by its leadership. By 2019, matters had deteriorated to an extent that DeepMind registered a new company called DeepMind Labs Limited, as well as a new holding company, according to filings with the UK's Companies House.

DeepMind has explored reorganizing itself as a 'global interest company' or a company limited by guarantee, a corporate structure without shareholders that is sometimes used by nonprofits. At issue is the concern that within the Alphabet structure its research will be used for military or surveillance purposes. Alarm bells rang through the company when reports emerged that Google had inked an agreement with the Pentagon, known as Project Maven, in 2017.

These fears weighed on DeepMind as far back as 2014 when, at the time of the acquisition, it sought an arrangement that would prevent Google from unilaterally taking control of its technology. According to an agreement signed between the two companies, called the Ethics and Safety Review Agreement, DeepMind's core AGI technology, whenever it is developed, will be placed under the control of a governing panel known as the Ethics Board and not directly put into the hands of Google. The company's CEO, Demis Hassabis, believes that AI technology should not be controlled by a single company. Speaking at the Tortoise Responsible AI Forum he thought out loud whether it would be better for it to be governed by a 'world institute' for AI which sits under the jurisdiction of the UN. 'It's much stronger if you lead by example,' he said, 'and I hope DeepMind can be part of that role-modelling for the industry.'

DeepMind's efforts to distance itself from its parent company came to nought in 2021, but reports suggest that it is still, at best, an uncomfortable union. In the summer of 2024, nearly 200 DeepMind employees signed a letter criticizing its parent company's involvement in military contracts at home and abroad. It noted: 'Any involvement with military and weapon manufacturing impacts our position as leaders in ethical and responsible AI, and goes against our mission statement and stated AI principles.' At its core this is a clash of cultures between an academically focused European research lab led by people who see themselves more as scientists than businessmen whose stated goal is to 'pave the way for truly beneficial and responsible AI', and a commercially minded

American corporate juggernaut which has products to ship and shareholder value to protect.

'If you are building a business in London or Paris or Amsterdam or Berlin, I think you are forced to think about value systems, not just, can I build it? Is it technically possible? Can I find assets? And then we'll figure the rest out afterwards – move fast and break things,' Saul continues. And that's the European challenge to the orthodoxy of how business is done in other places. 'I think what's happening is that we're breaking it down from this being about the monasteries in Silicon Valley and in Boston and maybe one or two other places to a reformation, a renaissance back to Europe.'

Saul's most contrarian bet as an investor might not be that the future of tech is in Europe or even that Europe is in fact 'one place', but that there can ever be a wildly successful tech company anywhere in the world that can achieve the improbable trifecta of being profitable, innovative and also seen by the broader public as somehow being 'good'. The amplitude of the task that confronts him is denominated not in mere billions but in trillions.

CHAPTER FOUR

Hyper Gap

'Our mindset is we need to be so far ahead that of our competition that it's a matter-of-fact thing to just give up because there's such a huge gap between the Korean companies and others.'

I

In 2010, Bom Kim, then a 32-year-old first-year student at Harvard Business School, reached out to Matthew Christensen, a friend who had previously worked with him at the Boston Consulting Group to ask if he would be willing to invest in his new startup. Christensen was less than enthused. He had already invested in Kim's previous venture, a magazine for Harvard alums called *12038*, named after the campus ZIP code, which in its marketing materials betrayed more than a hint of smugness when it declared that the periodical 'will deliver the world to our readers from the perspective they care about the most – their own'. Frequently referred to by Kim as the *Vanity Fair* of Harvard, the publication was a thinly veiled attempt at monetizing elite colleges' culture of self-congratulation, running breezy lifestyle pieces like a charmingly suggestive essay on dating titled 'Nowhere to Go But Down'. The venture got off to a promising enough start when it raised $4 million and was quickly bought by the publishers of *The Atlantic*. But banking solely on the preening self-regard of alums proved to be a less than sustainable business model and the magazine went bust within two years of its launch.

Undeterred by this setback, Kim, whose only previous entrepreneurial experience was in running college magazines, wanted in his next act to go even bigger. He arrived at the improbable ambition of going back to South Korea, a country he had left a quarter of a century earlier when he was only seven years old, to start a local variant of Groupon, a social ecommerce service that leveraged collective buying to get better deals for large groups of people which at the time was taking the US by storm. Christensen initially implored his friend to focus on his studies but eventually relented and his firm, Rose Park Advisors, which he ran with his father, the famed HBS professor Clayton Christensen, best known as the author of *Innovator's Dilemma*, a book that influenced Steve Jobs and is now considered a classic in management thinking, wrote a cheque for Kim's new company which he called Coupang, a mashup of 'coupon' and 'pang' – the Korean sound for hitting the jackpot.

Kim had hardly picked the path that was expected of him. Born in South Korea and raised in the US, he came of age thoroughly bathed in Americana. At 13, he enrolled in boarding school at Deerfield Academy in Massachusetts, where he lettered in varsity wrestling and track, then attended Harvard as an undergraduate and was well on his way to getting a second degree there when the lure of a rising Asia proved too strong to resist. He dropped out of business school after only six months to move to Seoul to work full time on his new startup, noting later that he had a 'very short window to really make something that had an impact'. It would prove to be a bumpy ride. The company had to reinvent itself three times before it hit its stride, going from being Korea's Groupon, to its eBay, to its Amazon. In 2012, Kim was forced to cancel the company's IPO a week before its listing because he realized he would have to completely revamp the business yet again before it stood a chance of making it big on the public markets.

What was at the time the most difficult decision Kim was confronted with as CEO has in retrospect proved to be the best call he ever made. When Coupang first launched it was the thirtieth odd startup to try and take on Korea's booming ecommerce market, the fifth largest in the world after China, US, UK and Japan. It has since edged out all of them to become by far the country's largest online retailer and among the ten largest in the world. Built in the image of Amazon – key hires are handed a copy of *The Everything Store* to read – the Korean upstart eclipses its American inspiration in important respects. Coupang runs its own logistics instead of relying on Korea's often rickety postal system and so can get items to customers faster than practically any other ecommerce platform in the world. Shoppers placing orders before midnight can expect their packages to arrive at their doorsteps as fast as 7 a.m. the next morning before they leave for work. Customers can return products by leaving them outside their door instead of having to deal with the hassle of boxes and return labels to mail them back. These conveniences are a boon for Koreans accustomed to working long hours in densely populated cities.

The company, which Kim has often compared to the Mongol empire, and whose mission is to make customers wonder 'how did I ever live without Coupang?', is now a ubiquitous presence in Korea and counts half of the country's 51-million people as its patrons with over two thirds of the entire Korean population living within ten minutes of a Coupang logistics centre. When it finally went public at the New York Stock Exchange on its second attempt in 2021, it debuted at a valuation of $84 billion, making it the biggest stock market listing in the US that year, the largest tech exit since Uber, and the second largest foreign IPO on Wall Street ever after Alibaba's mammoth showing in 2014. Rose Park Advisors, the low-profile investment firm run by the Harvard professor and his son which invested in Coupang at its inception on little more than a song and a prayer, saw its 5 per cent stake in the company score over $4 billion in returns. That was small change compared to Bom Kim's haul who saw his net worth balloon to over $10 billion, making the 42-year-old Harvard dropout and former intern of *The New Republic* the youngest self-made billionaire and the third richest person on the entire Korean peninsula.

2

The Coupang IPO was in every respect an international affair. One of Asia's largest companies, founded by an entrepreneur born in Seoul and raised in New England, with offices in Taiwan, Singapore, India and China had gone public in New York and in the process made billions for investors across continents from BlackRock in the US to SoftBank in Japan. What seemed natural or even inevitable in 2021 would have been unthinkable a few short decades ago in 1978 when Bom Kim was born. Back then Korea barely had enough dollars for itself let alone being able to make boatloads of greenbacks for investors overseas. It was ruled by a repressive military dictatorship which, in a bid to curb the flight of precious foreign exchange, did not permit its citizens to even leave its own borders. International travel was effectively prohibited and Koreans under the age of 50 were not allowed to have passports. Tourism was seen as frivolous luxury and generally considered out of the question. The lucky few who were allowed to travel did so only under special circumstances and only after submitting to an arduous bureaucratic process which involved attending anti-communism classes and depositing as surety a substantial sum of money in a national bank. Families were not allowed to go abroad together and travellers had to prove that at least some of their close relatives were staying back home. And even after all of that, they were issued a travel document that was typically valid for only a single use.

These restrictions were lifted as recently as 1989 after the country's transition to democracy. But even after that, it was not uncommon to see demonstrators picketing at Korean airports holding placards that read 'stop going overseas to play golf, waste dollars, and gamble'. Gambling is seen as a particularly egregious moral failing in Korean society and even today it is illegal for Korean citizens to gamble; the law is applied extraterritorially, meaning Koreans are subject to their country's gambling laws regardless of where they are in the world.

The 1970s were not that long ago. Microsoft and Apple had already launched their first products. The Rolling Stones were a thing. As was *Saturday Night Live*. They are all still around. And yet in this abbreviated time frame, while most of the world has changed incrementally, South Korea has gone through an exponential growth curve; moving swiftly from authoritarianism to democracy, middle-income to high-income, from having the same hermetic and ethnocentric tendencies as its neighbour to the north to becoming one of the most globalized places on earth. It is the world's fifth largest exporter of goods, most of them high-end technology products, like Samsung smartphones, Hyundai EVs and LG flatscreens and now increasingly cultural exports like the Netflix hit *Squid Games*, Academy Award Winner *Parasite*, K-pop, and blockbuster games like *PUBG*. The place that only recently fenced its citizens in so they couldn't leave now has a president who takes to the pages of *Foreign Affairs* to sell his country as the global pivotal state, or GPS, a self-conscious technological framing of its role in the world which has now become an oft-repeated three-letter shorthand meant to capture the foreign policy ambitions of a country which no longer sees itself as a 'shrimp among whales' but a technologically advanced middle power that can make its presence felt on the global stage. And all that change has happened from birth to midlife of one Bom Kim.

This transformation is even more striking when seen in the

broader sweep of the country's relatively brief history. Korea emerged as a modern nation state after the defeat of the Japanese in World War II. Prior to that, the entire Korean peninsula spent thirty-five years under brutal Japanese occupation which was sustained by a harrowing regime of forced labour and systematic suppression of local culture. Koreans were prohibited from speaking their own language and forced to adopt Japanese names. The end of colonialism brought hardly any immediate reprieve from suffering as the region was quickly thrust from the age of empire to the Cold War era of great power conflict. At the end of the war, the advancing Red Army seized the north and the US army took control of the south and the Korean peninsula, which had been a single entity through much of its history, was split at the 38th parallel north. The division, which was meant to be temporary, became permanent when the two rival great powers propped up their own governments in their respective zones: the Democratic People's Republic of Korea (DPRK) in the north and the Republic of Korea (RoK) in the south. The north, which at the time was stronger and more developed of the two, soon invaded its neighbour. The south almost totally capitulated. At the lowest point in the conflict almost 95 per cent of its territory, including its capital Seoul, was captured by enemy troops who forced the southern army to retreat to a small patch of land known as the Busan perimeter in the very southeastern corner of the country. It was only after military intervention by the UN, with a force comprising of troops drawn from twenty-one nations, with the US doing most of the heavy lifting, that the DPRK army, backed by the Soviet Union and China, was pushed back above the 38th parallel. The south, against all odds, held on to all of its territory but that was scarce consolation in the aftermath of a three-year armed conflict that wrought a humanitarian tragedy of catastrophic proportions. Over a million South Koreans were killed and half of the country's urban infrastructure and industrial capacity were destroyed. The

Republic of Korea emerged from four decades of colonialism and war as one of the poorest nations on earth with living standards comparable to those in the most economically disadvantaged parts of sub-Saharan Africa and below places like Haiti, Liberia and Yemen. The per capita income of North Korea was three times that of the South. The country was almost entirely dependent on foreign aid with three out of every four dollars spent by the government coming from overseas. An average Korean made less than $100 a year and in the early years it was not unusual to see hunger-stricken people scouring rocky mountains in search of edible herbs and plants to eat.

Today, South Korea is firmly a member of the developed world with standards of living comparable to those in Italy or Spain. Its per capita income ranks ahead of its erstwhile colonizer Japan, and worker productivity is three times that of China. It is one of the world's ten largest economies, a member of both the OECD and the G20. The country that one lifetime ago was almost entirely dependent on overseas assistance is now among the world's twenty largest foreign aid donors, giving out almost $4 billion a year, and is on track to be one of the ten largest within this decade. This remarkable transformation of this war-torn country into an economic, technological, and now increasingly cultural powerhouse, now known as the 'Miracle on the Han River', has become an oft-cited case study in national development taught in economics courses around the world as one of regrettably few success stories of a formerly impoverished and colonized nation making the leap from authoritarianism to democracy and poverty to prosperity all in the span of one generation.

3

Korea's economic miracle is the work of sustained efforts by successive governments. But one man casts a longer shadow than others over the country's post-colonial history. Park Chung-hee, a high-ranking military official, seized power in a bloodless coup in 1961 and ruled the country under strict authoritarian control for an unbroken stretch of eighteen years until he was assassinated by his own intelligence chief in 1979. Park's rallying cry was 'pukuk kangbyong', or 'rich country, strong army', a slogan that captured the young country's overriding concern with staving off the existential threat from the north which had come as close as it possibly could to wiping it out at its very inception. Park's military junta went about revamping the economy through measures that would become standard practice in the developing world: heavy state intervention aimed at curbing imports, promoting exports and nurturing domestic industry by shielding it from foreign competition.

But the regime's close relationship with the country's private sector, which would set Korea down the path of supercharged development for which it is known today, got off to a rocky start. In the early years, Park forcibly nationalized banks and rounded up all the prominent businessmen and threw them in jail, some of whom were paraded through the streets in dunce caps with placards that read: 'I am a corrupt swine.' Even Lee Byung-chul, the founder of Samsung and the country's richest man, was not spared. The oligarchs managed to bargain for their release by promising to rebuild the country in return for state support with Byung-chul famously declaring that 'government and industry are like husband and wife'. This alliance spurred the growth of the chaebols, South Korea's powerful industrial conglomerates like Samsung, LG, Hyundai, Lotte and SK which still loom large over its economy and are the face of corporate Korea abroad even today.

Chaebol, which is a combination of the word *chae*, which means wealth or money, and *bol*, which means clan or clique, so literally money-clique, is a corporate structure unique to Korea. These sprawling family owned businesses operate across multiple unrelated sectors – the same company can be in banking, real estate, transport, electronics, hospitality, healthcare and dozens of other industries – and were built with heavy government assistance that came in the form of subsidies, loans and tax incentives. There are more than forty conglomerates that fit this description but just a handful wield enormous economic clout: the top ten hold over a quarter of all business assets and the top five account for over half of the entire value of the South Korean stock market. The family dynasties that run these businesses are among the richest in the world and enjoy celebrity status in Korea whose foibles are followed with as much enthusiasm as those of Elon Musk and Kylie Jenner in the US. These families have sustained and even extended their influence over national affairs even as the relevance of other institutions, like the previously all-powerful military, has faded.

The chaebols have modest origins and many of them got their start selling textiles, plywood, and even wigs. That's no joke, South Korea has a long history of cornering the global market for hairpieces and this obscure industry played an outsized role in the country's early development. In the 1960s, wigs were the country's third largest export, earning one in every ten dollars it received in foreign exchange, with up to a third of the wigs worn in the US coming from this small-ish country in the Far East. Over time these conglomerates entered manufacturing, heavy industry, and then eventually electronics and high-tech sectors where they now punch at the highest level globally. Hyundai, which started as a small construction company, is the third largest automaker in the world and recently became the second most popular EV brand in the US after Tesla, ahead of the nation's oldest car companies like Ford and GM. One in every ten EVs sold in the US is a Hyundai. SK Hynix is the world's leading producer of high bandwidth memory chips which are crucial components of Nvidia's graphic processing units, the infrastructure backbone of virtually all major AI applications. These chaebols, which in their early incarnations operated in sectors considered too frivolous to matter, are now among the largest companies in the world. Thirteen of the world's largest 500 corporations are Korean. Some have reached a scale larger than most national economies. The combined revenues of just the top ten chaebols, which top a trillion dollars, are larger than the GDP of all but the twenty largest economies in the world. It's these family owned conglomerates that have been the engines of Korea's remarkable ascent, with *The Atlantic* noting that 'without them, South Korea would look more like emerging Vietnam than developed Japan, the only other Asian country to join the Organization for Economic Cooperation and Development's club of "developed" economies.'

4

A handful of chaebols have grown to be among the world's largest companies but even among these behemoths one stands out. The Samsung Group, whose numerous subsidiaries have a combined value of over half a trillion dollars, is at least five times larger than SK Hynix, its second ranked competitor. Samsung is among the world's fifteen largest corporations and Asia's third largest after Aramco and TSMC. To the outside world Samsung is known mostly for electronics and home appliances: smartphones, televisions, washing machines and refrigerators. But in Korea the company's presence is much bigger, almost all-encompassing.

Samsung runs more than eighty distinct businesses, and it does everything from building roads and ships and oil rigs to owning hospitals and theme parks and sports teams to running entire cities. There is an old joke that goes that in Korea you can live an all-Samsung life from cradle to grave: a person can be born in a Samsung-run medical centre, grow up in a Samsung-owned apartment fitted out entirely with Samsung appliances and built by a Samsung-owned construction firm in a city built from scratch by

Samsung, where they wear Samsung-branded clothes, study in a Samsung-run school, and then, when they enlist in mandatory military service, find themselves in a Samsung-made tank, then go to work in one of virtually hundreds of Samsung-run businesses, to be admitted to which they had to of course take the Global Samsung Aptitude Test, or GSAT, then get married in a Samsung marriage bureau, and retire into a Samsung retirement home and eventually end up at a Samsung-run funeral parlour, all of which brings us that much closer to a world where it's not entirely out of place to run an advert during a eulogy: 'Today we engage in celebration of the life of Aunt Marianne, brought to you by Samsung.'

No other private company in the world plays as expansive a role in its domestic economy as Samsung does in Korea. This one company, which the *Washington Post* once called a 'do-everything monolith', is responsible for a full fifth of all exports and a fifth of Korea's entire GDP, which is remarkable given Samsung, like other chaebols, is not a purely public company but at its core is still a family owned conglomerate in which the descendants of its founder, Lee Byung-chul, call the shots on virtually every decision that matters. So pervasive is Samsung's presence in everyday life that some Koreans often cynically refer to their country as the 'Republic of Samsung' and even a level-headed assessment of the company's sway over national affairs would place its influence as being second only to that of the Korean government. The company, which has in recent years fallen foul of public opinion, was through much of its history seen as more than just a business and equated with the nation itself, a central plotline in Korea's transformation from a war-ravaged nation to the fourth largest economy in Asia.

Even beyond Korea's borders there's hardly any other company that can match the sheer breadth of Samsung's technical prowess and the global scale of its operations. This is the company that was a central player in the consortia that built the world's tallest building, the Burj Khalifa, in Dubai, as well as the second tallest building, the Merdeka 118, in Kuala Lumpur, and what was at the time of its launch the world's largest container ship, the MSC Gülsün, registered in Panama, and the Barakah nuclear power plant in the UAE, the first in the Arab world, which provides a quarter of the entire country's electricity, as well as the Prelude FLNG, a floating liquified natural gas platform, owned by Shell and moored off the coast of Western Australia, which is the world's largest floating object, six times larger than the biggest US aircraft carrier, a vessel that can hold enough fuel at a time to heat all of London for a week and is strong enough to withstand a category 5 cyclone.

This colossus of a company has decidedly modest origins. Samsung, which literally translates to three stars, an early sign of the company's extravagant, world-conquering ambitions, was founded in 1938 when Korea was still under Japanese colonial rule as a small trading company with only $25 in capital which sold noodles and dried fish. After independence it expanded into sugar, textiles, insurance and retail and it wasn't until the 1960s that it made its first foray into electronics with the launch of a black and white television. Its founder's vision was to build a company in the image of Mitsubishi and in the early years the company behaved like a traditional Japanese zaibatsu which enforced rigid military-like discipline and executives did as they were told without argument. And so, in its first incarnation, which lasted over half a century, Samsung was seen as little more than a manufacturer of cheap knockoffs of everyday consumer electronics – microwaves, VCRs, video cameras – that had already been introduced in Western markets, 'Sam-suck' being a common refrain to those pondering taking a dip into buying its products.

It wasn't until the 1990s after the company's founder passed away and a new generation took over that its reputation began to change. In 1993 the new chairman, the founder's third eldest son, Lee Kun-hee, took a global tour across the sprawling empire he had inherited, ending his trip in Frankfurt where he assembled his top lieutenants in a meeting that lasted three days where he exhorted his management to 'change everything but your wife and children'. The event, which within Samsung became known as the Frankfurt Declaration, marked the company's pivot from a second-tier, high-volume, low-quality manufacturer to a high-end design powerhouse which would in short order become the biggest and most powerful electronics company on earth.

Nowhere is Samsung's dominance in hardware more visible than in what is perhaps the most consequential and ubiquitous technology platform launched in the past quarter of a century: smartphones. It is the world's most popular smartphone manufacturer, ranking ahead of Apple, Xiaomi and virtually hundreds of other cellphone makers in what is arguably the most competitive market for any technology product globally. One in four people who use a smartphone use a Samsung. The company's ascendancy in smartphones is even more striking given the fact that it was the only major cellphone manufacturer that survived the seismic shift that occurred in the industry with the launch of the iPhone.

Samsung has been making handhelds since the 1980s and after spending decades in the shadows of Scandinavian and Japanese competitors was able to claw its way to second spot behind Nokia which is where it ranked when Steve Jobs took to the stage at Macworld in early 2007 to launch his revolutionary device. And yet even though virtually none of the giants of the pre-smartphone era – Nokia, Ericsson, BlackBerry, Motorola – were able to stay relevant with the shift in technology platforms, Samsung not only adapted, but thrived, outpacing Apple within half a decade to take the top spot as the world's bestselling smartphone manufacturer with the *Wall Street Journal* asking 'Has Apple Lost Its Cool to Samsung?' as far back as 2013. In doing so, the company made itself a rare exception to the *Innovator's Dilemma*, a term coined by Clayton Christensen, the Harvard Business School professor who was an early investor in Coupang, and refers to what has been a consistent trend of a dominant technology pioneer of one era completely missing the boat when confronted with the next big wave: Kodak and digital photography, Blockbuster and video streaming, Intel and GPUs being the most commonly cited examples. Samsung has repeated this feat in product category after product category, from TVs to batteries to semiconductors to memory chips.

Samsung's presence in the smartphone business is a lot more pervasive than its sales figures would suggest. It is also a major supplier of components to its main competitor, Apple, with the two behemoths between them shipping almost half of all smartphones sold globally every year. The relationship began in 1983 when a 28-year-old Steve Jobs met Lee Byung-chul to source memory chips for a tablet computer that he was planning – a full twenty-seven years before the launch of the iPad. It would be the beginning of a long and complicated relationship with the South Korean giant providing semiconductors, memory chips and display screens for Apple devices, landing it in the unusual and enviable position of making money not only when it sells a smartphone or tablet itself but also when its main competitor sells one as well. It has been an uneasy alliance, with the two tech giants, often described as frenemies, the most dominant players in the long-running rivalry between iOS and Android, a headline battle in tech, frequently suing each other for patent infringement, while also over time deepening a relationship which has seen Samsung become Apple's biggest supplier and Apple in turn become Samsung's single largest corporate customer.

5

It used to be considered a truism that what is good for Samsung is good for Korea, with the chaebols widely credited with being the catalyst of the country's economic transformation. But this consensus has in recent decades largely broken down. The economic model that was the source of Korea's postwar turnaround is now facing popular backlash for concentrating too much economic power in too few corporations, with critics charging that the tenth largest economy in the world has effectively become a playground for a handful of families. Two out of three Koreans think that the chaebols need to be reformed and support for such measures is even higher among young people with three out of every four of those in their twenties and thirties in favour of reining in the country's flagship conglomerates. There is a growing consensus that if political democratization was the last major milestone in Korean history, then economic democratization ought to be the logical next step in the evolution of its society.

The conglomerates don't just play an outsized role in the economy. Their influence also spills over into politics. This sometimes happens in very overt and legitimate ways, like when Lee Myung-bak, a former CEO of the engineering arm of Hyundai, was elected the country's president in 2008. But this cozy relationship between government and industry mostly plays out away from the public eye in more shadowy arrangements. Until recently it was considered quite normal for politicians to ask chaebol executives for money in return for political favours. Korean corporate history is replete with scandals implicating senior chaebol figures in every white-collar crime imaginable, from tax evasion to bribery to embezzlement. Most are handed light sentences and some let off the hook entirely. Courts have a long history of judicial leniency towards chaebol executives, who are often deemed 'too big to jail' on account of fears that penalizing them too harshly can have negative 'system risk' for the economy. A familiar pattern has emerged in the sentencing of high-profile financial crimes in what is often called the 'three-five rule' which has become the exclusive privilege of senior chaebol figures: a guilty executive is handed a three-year prison sentence which is suspended for five years and then eventually commuted so they effectively serve no prison time at all.

Lee Kun-hee, former Samsung chairman and son of the conglomerate's founder, the man who issued the famous Frankfurt Declaration which changed the direction of the company, was convicted twice, once for bribery and once for tax evasion, only to be pardoned on both occasions. One of these cases also ensnared Lee Myung-bak, the former Hyundai executive, who was accused of taking $6 million from the Samsung chairman in exchange for the pardon. The former president was handed a fifteen-year prison sentence but was also eventually pardoned. This case was seen as Exhibit A for how deeply entrenched chaebol influence is in Korean society with the principal extending the bribe and the one collecting it, the one offering the pardon and the one receiving it, both being boldface names of the chaebol establishment.

Resentment at the growing influence of a handful of companies, which had been on slow burn for decades, boiled over into a full-blown public crisis in 2016 when Lee Myung-bak's successor to the presidency, Park Geun-hye, was also found to have inappropriate financial dealings with the chaebols. The election of Park Geun-hye, the daughter of Park Chung-hee, the military dictator who is considered the father of modern Korea, was seen as a significant political milestone for a country which, in another lightning fast triumph in its abbreviated history, had managed to elect a female leader within twenty-five years of its transition to representative government, a feat which has proven elusive for far more established democracies. But the elation was short-lived. Park was brought down in a bizarre Tolkienesque scandal after it was revealed that she was effectively controlled by her spiritual advisor, Choi Soon-sil, a woman who held no official government position but had nevertheless managed to accede some of the powers of the presidency.

The advisor had inherited her hold over the president from her father, a shaman who had set up a religious cult called the 'Future

Life Church' and declared himself to be the 'Future Buddha'. The US Embassy in Seoul had once described him in a secret cable as the 'Korean Rasputin' who had 'complete control over Park's body and soul during her formative years'. He had first made contact with the future president when she was still a young woman, claiming he could see visions of her deceased mother, the former first lady who was assassinated in 1974 in a misfired shot aimed at her husband, asking him to help her child. He soon became the young Park's mentor and so began a decades long relationship in which the shaman and then her daughter would deliver messages from the afterlife to the offspring of the most consequential figure in modern Korean history who herself would go on to become the first female to hold the country's highest political office.

Unsurprisingly, there was a lot of money involved. The advisor, Choi Soon-sil, was found to have been running a racket and had extorted over $60 million from the chaebols, funnelling these funds into her foundations. Here too Samsung was implicated. The conglomerate had given over $35 million to Choi's non-profits and other donations which included a million-dollar horse, named Vitana V, gifted to her equestrian daughter who lived in Europe, with prosecutors alleging that the donations were made to clear regulatory hurdles for a proposed merger of two Samsung business units, a characterization the company disputes, presenting itself instead as a victim of Choi Soon-sil's shakedown. The revelations sparked mass demonstrations that brought 16 million people, nearly a third of the country's entire population, into the streets over twenty consecutive Saturday nights to demand the resignation of the president. The protests, the largest in Korean history, came to be known as the Candlelight Revolution and eventually forced Park Geun-hye from office, who was then impeached and sentenced to twenty-four years in prison. Lee Jae-yong, the de facto head of Samsung, was also sentenced to five years in prison. Both were pardoned before serving out their full sentences.

Critics charge that the culture of impunity for the powerful has effectively created a two-tiered society in which one set of rules apply to the chaebols and their associates and another to everyone else. Reminders of this stratification don't always come in the form of outright criminality but often as scandals involving senior business figures behaving badly, which happens so frequently that there's even a term for it, *gabjil*, which roughly translates to 'high handedness'. The most infamous incident occurred in 2014 when Heather Cho, 40-year-old heiress of the Hanjin Group, a major chaebol, was seated in first class on a flight from New York to Seoul on Korean Air, which her family owns, when a flight attendant served her macadamia nuts in a bag instead of on a plate as was her preference. Enraged by the insolence she, according to eyewitnesses, 'snarled like a wild beast', and made the attendant get on his knees and ask for forgiveness, while repeatedly striking his knuckles with her tablet. She then proceeded to order the pilot to return the plane, which was already making its way to the runway, back to the gate so that the offending crew member could be ejected from the aircraft. The episode, now immortalized as the 'nut rage' incident, caused a national furore resulting in a criminal conviction for Heather Cho who was given a one-year prison sentence. She was released in five months. Other high-profile incidents of *gabjil* have included a chaebol owner assaulting a security guard because he tried to close the restaurant before he finished dining and another who kidnapped youths who had got into a fight with his son at a bar.

Koreans might have been more willing to overlook these foibles if they thought that the chaebols were delivering on their original promise of strong business performance and wealth creation, which in turn served as a solid economic foundation for shared prosperity. But that narrative too is falling apart. The economic contribution of these conglomerates, which was indisputable in the first half century after independence, is now seen by many as a net liability.

The watershed moment for this came during the 1997 Asian Financial Crisis, the most traumatic experience the country went through since the Korean War, during which its currency lost over 75 per cent of its value against the dollar. The government had to go hat in hand to the International Monetary Fund – a humiliating experience for a country which had in the years in the run up to the crisis been feted globally as a resounding economic success and begun to see itself as firmly a member of the rich world – and was offered a bailout package of $58 billion, the largest ever in history. The loan saved the country from immediate economic catastrophe but not without inflicting deep psychological scars. To raise funds to pay the debt, the government appealed to regular Koreans to voluntarily hand over their savings and they responded with a level of patriotism rarely seen anywhere in the world outside of a war. Nearly 3.5 million people, or a quarter of the country's population, queued up outside donation points to hand over household gold: rings, necklaces and all manner of trinkets. Lee Jong-beom, a baseball star, handed over all the trophies and medals he had won in his career. In two months, 226 tons of gold, worth over $2.2 billion, were collected, all of which was melted and sent to the IMF.

The causes of the crisis were complex but there is consensus that it was in part precipitated by the declining competitiveness of Korean industry, which for decades had expanded on a steady supply of cheap credit assured by its close ties to the government, which privileged cozy connections over hard-headed credit analyses and risk assessments. The chaebols, thinking they were too big to fail and would always be bailed out by the state, took on dangerous levels of debt, much of which was denominated in dollars and issued by foreign creditors. When the value of the currency collapsed, many of these debt-fuelled businesses could no longer fulfil their financial obligations and went bankrupt. These included Daewoo, the country's second largest chaebol, which, when it imploded with a $50 billion hole in its accounts, became the world's largest bankruptcy in history until Enron went under in 2001. Korea eventually paid back its international lenders in full, three years ahead of schedule, and resumed its breakneck ascent. But it took a lot of pain and shared sacrifice to get there. It emerged out of the crises having learned some hard lessons, none bigger than the need to recalibrate its relationship with the chaebols: the business groups that were hitherto seen as the economy's indisputable champions would from now on be increasingly cast as its principal villains.

6

The charge against the chaebols wasn't just that they were complicit in the Asian Financial Crisis but that these monopolies were preventing a newer generation of companies that could put the economy on a more diversified and stable footing from coming to the fore. 'In the old economy you had to be a supplier to Samsung or you work for Samsung, that was a major, major thing,' Jimmy Kim, the co-founder of SparkLabs, a startup incubator in Seoul, told me. 'You're the first vendor, second vendor, third vendor, you're in that ecosystem, that's how you survive.' The AFC prompted a major rethink on how the economy ought to be organized and had a cascading effect on the country's longstanding political order. The national election on the heels of the crisis in 1997 marked the first time in Korea's post-independence history that an opposition candidate won the presidency with a peaceful transition of power from the ruling party to the opposition. In his inaugural address, the new president, Kim Dae-jung, directly called out the power of the conglomerates by noting that the crisis 'would not have taken place if the political, economic and financial leaders of this country were not tainted by a collusive link between politics and business'. His answer to the crisis was specific and progressive for that time: state backing for a new generation of venture companies. 'Venture backed companies are the flower of the new century,' he noted in his inaugural address, adding that 'we will resolutely push a policy to make our nation strong in leading-edge technologies'.

The timing of this shift in economic priorities was fortuitous in that it coincided with the rise of the internet and the large-platform companies that came with it. The government saw an opening to build a new digital economy away from chaebol control. It made funding available for new companies and within three years up to a third of the venture funding came directly from government sources. It also offered loans to new startups at lower than market rates and eased the requirements for listing on the Kosdaq, the Korean Nasdaq, which made it easier for new companies to raise money from public markets. It simultaneously imposed restrictions on chaebols from getting too deeply involved in the new sectors – they could invest in new ventures but not own more than 30 per cent of the business and that too if they were not the largest shareholder.

The growth of new companies after the AFC was also helped by talent leaving the older conglomerates at unprecedented rates. Prior to the crisis there was an unwritten social contract between the chaebols and their employees that they would have a job for life. But with these conglomerates under financial pressure like never before, this contract fell apart and in an unprecedented move thousands of employees were let go and many left voluntarily. This recycling of talent would prove to be beneficial to the Korean economy as some would go off and seed the next generation of Korean enterprise. Kim Beom-su and Lee Hae-jin, who first became friends while working at Samsung, would both leave to work on their own internet businesses, which they eventually merged to form Korea's largest internet portal, but their relationship turned competitive and the two went their separate ways and became rivals. Lee Hae-jin would lead Naver, a search engine, like Korea's

Google, and Kim Beom-su would lead Kakao, which first came up as a messaging service, like Korea's WhatsApp, and later became an app for everything. Both businesses would grow to the scale of tens of billions of dollars and by 2021 had broken into the ranks of the top five highest valued companies on the stock market, turning their respective founders into self-made billionaires, until then a rare feat in chaebol-dominated Korea.

At first Kakao and Naver were seen as heralding a new era for Korean business. Government policy had been successful: the winners of the new digital economy were not the same as the winners of the old industrial economy and it seemed like the stranglehold of the chaebols over industry had finally been broken. The story of Kakao and its founder Kim Beom-su resonated particularly strongly with the public coming as he did from a background that seemed to be at odds with the mainstream of the old business establishment. Kim was born into a family of modest means: his father worked in a pen factory and his mother was a maid in a hotel. He grew up in one of the most disadvantaged neighbourhoods of Seoul where he shared a one-bedroom apartment with eight other family members. He was the first in his family to make it to college when he was admitted to Seoul National University, Korea's equivalent of Harvard, where he paid his way by tutoring on the side. After graduating with a degree in engineering, he landed a job with Samsung, the most eminently respectable thing a college graduate could do in Korea at the time. He would eventually become an entrepreneur and move to the US where in 2007 he came across the first iPhone and instantly saw its commercial potential. He would soon return to Korea and in 2010 launch KakaoTalk, inspired by WhatsApp, which would quickly become the most popular messaging service in the country.

The company's influence on the digital economy spilled over into the rest of Asia. Kim has longstanding partnerships with Tencent and Naver and both relationships would eventually turn competitive. Tencent adopted some of the features of KakaoTalk into its own superapp, WeChat, while Naver would launch a similar service in Japan, called Line, and both of these rival services would become the biggest messaging services in their respective countries, much larger by comparison, leaving Kim with the consolation prize of the much smaller Korean market and depriving him of the opportunity of building a truly international brand. It's hard to feel too bad for him though. Kakao would go on to become one of the three most valuable listed companies in Korea, and when the online economy took off during the pandemic Kim Beom-su pulled ahead of Jay Y. Lee, an heir to the Samsung dynasty and the chairman of the country's biggest company, to become the richest person in Korea. This was seen as a watershed moment in Korean industry, a symbolic passing of the baton from the previous generation of family owned businesses run by heirs of inherited fortunes to the newer generation of public companies built by self-made tech billionaires. Kim Beom-su consolidated his place as the face of responsible capitalism in Korea when he joined the likes of Bill Gates and Warren Buffett to sign the Giving Pledge, a commitment to donate a majority of his fortune to society, only the second Korean ever to make that pledge.

Life is messy, things rarely work out as expected, and this is where this Cinderella story takes a hard right turn into despair. Korea's infatuation with the winners of its platform economy would be short-lived. It didn't take long for the newer platform companies to start engaging in patterns of behaviour starkly reminiscent of the old chaebols: privileging family ties over professional management, creating sprawling empires that spanned vastly distant industries, and engaging in anti-competitive behaviour. The distinction between ownership and management, which is the hallmark of public companies in more mature economies, became with these platforms progressively blurrier. At one point Kim had hired his wife, brother and two children as company executives. If the charge against the chaebols was that they crowded out newer companies by becoming do-everything monoliths that used their market power to create walled gardens around their customers so they could only use their services to the exclusion of everyone else, then the newer platform companies too were becoming all-encompassing digital ecosystems. By 2023, Kakao had spawned 175 subsidiaries, more than any other company apart from the SK Group, a chaebol. It has a presence in everything from messaging to gaming, banking, payments, ride-sharing, digital publishing and search, to name just a few. The *Washington Post* would describe Kakao as 'Facebook messenger, WhatsApp, Uber, Google Maps and Venmo wrapped into one'. Within a decade of its founding this one company had become so central to mediating the everyday lives of Koreans that when one of its data centres caught fire in December 2022 the disruptions were felt across the nation, with everything from small shops to online businesses to taxi drivers being unable to process payments. Even the country's president had to weigh-in on the disaster with an acknowledgement that he considered Kakao's services to be a part of the 'basic infrastructure' of the country, a characterization that was often used in the past for the too big to fail chaebols. 'Kakao has turned from a symbol of growth and innovation into a symbol of old greed,' Song Young-gil, a representative of the ruling party, told the National Assembly in 2021. 'We will find a way to stop its rapid expansion.'

Kakao now finds itself tangled in regulatory issues that have long been part of the cost of doing business for just about every chaebol. In 2023, Korean prosecutors brought charges against its founder and senior executives for engaging in anti-competitive behaviour. The case centred on the alleged manipulation of the stock price of SM Entertainment, a K-pop firm it wanted to acquire in a hotly contested battle with HYBE, the firm that manages the boyband BTS. Prosecutors accused Kim of artificially inflating SM Entertainment's share price in a bid to block its takeover by HYBE.

In the summer of 2024, Kim was arrested in connection with this case, which brought things full circle, with the man who was once thought to be the challenger to the chaebol-dominated order now meeting the same fate as that of the old corporate bosses. It also signalled a reversion to mean. Kakao and Kim's fortune plummeted and the country's rich list was once again populated by boldface names of the chaebol establishment. The episode throws open fundamental questions about whether the original prognosis of Korea's economic woes that emerged in the aftermath of the Asian Financial Crisis was the correct one. The government had succeeded in its policy to shift the economic centre of gravity away from the conglomerates. And yet simply changing the form of business organization did not address the fundamental tendency toward economic concentration in the country's corporate ecosystem. Perhaps the problem lies not in the chaebol structure itself but in deeper institutional and cultural factors that continue to enable market dominance by a few powerful entities. And if so, then how tractable is that problem really to government policy?

7

The story of Korean tech is usually told in three acts. First came the big chaebols like Samsung and Hyundai. These were large industrial firms that initially made cheap knockoffs and then eventually got into more high-end manufacturing. Then came the big internet platform companies like Naver and Kakao. These were very similar to companies that already existed elsewhere – Google, WhatsApp – but modified their products for a captive domestic market and that's how they got to scale. The first generation of companies was global, they are still the face of Korean business in the world. The second generation of companies would not be that familiar to people outside of Korea. But far from being a sign of the waning competitiveness of Korean business, it's an indication of the opposite: the growing strength of the Korean economy. The first generation of Korean businesses had to focus on exports because the country just didn't have enough wealth going around for it to be an attractive market. They had to earn foreign exchange abroad. But with the rising tide of the Korean economy, its companies can now serve their own people and still grow to the size of the largest businesses in the world. That's the story of Coupang, one of the biggest stock market debuts in the US, which got there even though its customer base is entirely Korean.

By most accounts Korea is now in the third generation of tech companies. These would be names like Krafton, the maker of the blockbuster *PUBG*, which would be well known to gamers everywhere, and then other startups like Sendbird, TwelveLabs, Moloco, GeneEdit and Lunit which are worth billions of dollars but are still not all that visible abroad. In their case it's because they make software for other businesses rather than selling products and services to regular people, so by the very nature of their business they are inherently low profile. I spoke to their founders to find out what the newer generation tech businesses are all about and how they see what's happening in tech in Korea. Two things immediately stood out. The first was that even if all these companies are founded by Korean entrepreneurs they are best understood as hybrid companies since their presence spans both the US and Korea. My impression was that having a dual presence of one foot in the US and one in Korea was considered aspirational by the new generation of Korean entrepreneurs. The second was that their businesses were not about copying what was already available elsewhere, the model of the first generation of Korean companies, or relying solely on monetizing a captive domestic market, the model of the second generation, but on real technological advantage over their competitors. They are at or ahead of the curve in frontier disciplines like using artificial intelligence for cancer detection and developing novel gene-based therapies, often at the forefront of the state of the art in the world.

Jae Lee grew up in the US and was among the first Koreans to attend Exeter, a well-known boarding school in New Hampshire. After graduating from Berkeley with a degree in computer science in 2017, he went back to Korea to enrol in mandatory military service where instead of being handed a rifle he was sent to Korean Cyber Command. His job was to work on indexing the dark web to make it searchable specifically for undesirable video content. It was here that he met the co-founders for his future startup. They eventually left to launch TwelveLabs, which uses artificial intelligence for video search. Search is obviously big business, it has spawned the biggest companies of the digital era, but its capabilities have so far been limited to text and photos. Video search is notoriously hard to crack and TwelveLabs is betting that they can solve this problem. Some of the uses of video search are obvious: going through large video libraries to find the right frame or skipping ahead on a movie or a YouTube video to just the information you need. Other uses are less obvious but no less lucrative, like contextual advertising, which can change the trillion-dollar global advertising industry. The current state of the art in advertising is to present targeted ads that are tailormade to the audience profile and viewing habits. With contextual advertising the algorithms will be able to detect what you are watching then present an ad that is directly related to the content of your video: like say you're watching a movie that has a scene from Venice, TwelveLabs' algorithms can detect that and present you with an ad for travel tours to Italy. Jae, who splits his time between San Francisco and Seoul, tells me that much of what he is working on is in the spirit of the hyper-gap, a government sponsored drive to elevate Korean business to a different plane from its competitors. 'Our mindset is we need to be so far ahead of our competitors that it's a matter-of-fact thing to just give up because there's such a huge gap between Korean companies and others,' he says.

Among the more well-known of the new generation of Korean

entrepreneurs is John Kim, who was the number one ranked Unreal Tournament gamer in Korea before he turned his energies to building companies. He says that the same methodical approach that helped him climb the ranks of the gaming world has also held him in good stead as an entrepreneur: 'I created routines as a system to improve myself in a very deliberate way. How do I get better at shooting rail guns? How do I get better at shooting at rocket launchers? And how do I improve my accuracy versus speed? It's kind of a divide and conquer engineering mindset.' Kim's company Sendbird, now worth over a billion dollars, helps companies develop communication features in their web and mobile applications. The chat features in apps like Reddit and Hinge are the work of Sendbird. Kim, who is a Y Combinator alum, says that while Korea has made substantial progress in developing its entrepreneurial culture in the past decade, other challenges have emerged, like the country's rapidly collapsing population. Korea has the world's lowest birth rate, which dropped to 0.72 in 2023, far below the replacement level of 2.1 needed to maintain a stable population. 'No company can beat demographics when it comes to market size,' he says.

Entrepreneurs running this newer generation of hybrid companies benefit from the strong relationship that Korea has enjoyed with the US since its founding, steeped as it was in the politics of the Cold War. These ties have only grown since. Korea sends the third largest cohort of international students to the US, behind only India and China, which is impressive given that they have over a billion people each. Relative to its population Korea sends the greatest number of international students to the US of any country in the world. The country also benefits from the large Korean diaspora in the US, some of whom, like Bom Kim of Coupang, are reconnecting with the motherland given its recent turn in fortunes.

Han Kim was born in Korea but moved with his family to the

US when he was only ten years old. He attended West Point and then served as a lieutenant in the US Army, a role that took him back to Korea where he commanded Korean troops which rekindled his relationship with the country. A child of the Cold War, he looked for new meaning in life after the fall of the Berlin Wall which took him to business school at Stanford. After his MBA he launched Altos, a venture capital fund based out of Silicon Valley which came up in 1996. In its initial years, the firm struggled to break into an industry dominated by older firms with well-established local networks, and after a long dark period survived by diversifying out of the Valley to other cities in the US overlooked by the mainstream VC industry, and then eventually also expanding into South Korea. Altos would go on to become one of the few firms to do well on both sides of the Pacific. It made an early bet on Roblox, a gaming company, which, when it made its stock market debut in 2021 at a valuation of $41 billion, netted Altos, its largest shareholder, over $8 billion in returns. It also had a winning streak in Korea with early investments in the biggest names of the new wave of Korea tech: Coupang, Woowa Brothers and Krafton, establishing itself as a top-tier VC fund in the country. 'I do think the Korean startup community has benefited from Korean-Americans that came into Korea,' Han tells me. 'The ones that meshed well with the Korean Koreans I think have succeeded. The ones that sort of had the attitude that I know better, they all did not do well.'

Another Korean-American tech executive, who followed the well-worn path of an Ivy League education followed by a career on Wall Street only to go back to Seoul, told me that a rising Korea presented him with an opportunity set that was a lot more compelling to what lay ahead of him in the US. 'As a 20-year-old coming onto Wall Street back in 2009 there really weren't as many role models that you could look up to that were Korean who had sort of broken through the glass ceiling,' he told me. 'I realized that if I stayed in investment banking I would still sort of be a cog in a wheel at 45, I'll be a managing director, if I play my cards right, play a lot of office politics, but I'll be a managing director covering a sector or covering a country like Korea, which is a relatively small country from a global investment allocation perspective, and then you have a decent family of two kids and live in Westchester. It's just boring really.' He added: 'I just think there's a lot more interesting opportunities when you come back to a relatively small country, you can do a lot more earlier in your life.'

Even the entrepreneurs who make the trade the other way around, moving from Korea to the US, have found it beneficial to maintain a relationship with the tech community back home. Kunwoo Lee got his undergraduate degree at the Korean Advanced Institute for Science and Technology (KAIST) before moving to the US to do his PhD in biomedical engineering at UC Berkeley where he worked closely with Jennifer Doudna, the Nobel Prize-winning pioneer of CRISPR technology. CRISPR (Clustered Regularly Interspaced Short Palindromic Repeats) is a gene editing tool which works like molecular scissors that can precisely cut DNA at specific locations, allowing scientists to modify, delete, or insert genes with precision. This opens the door to a whole new era of medicine. Even though CRISPR can edit genes precisely, it still needs to reach the target cells in the body to work, and that is where Kunwoo's company comes in. GenEdit develops delivery systems that can take CRISPR components where they are needed. Although the company has a strong US-centric identity, it also maintains a presence in Korea. 'Building a lab and manufacturing costs are nine times lower in Korea,' Kunwoo tells me. 'We decided that there is no reason not to leverage that and we opened an additional site in Korea.'

Much of this manufacturing capacity in Korea is the work of Samsung, which in 2011 launched a biotech division, called Samsung Biologics, which quickly became a world leader in biopharmaceutical manufacturing. Which brings us full circle to the enduring role of the chaebols in the Korean economy. Unlike other countries where the older generation of tech companies have been completely replaced by what came after – think IBM, Nortel, Nokia – in Korea these older companies sit alongside the later generations and are still aggressively entering new sectors in which they often punch at the highest level globally. The newer firms are promising but their value still lies mostly in the future. When it comes to the demonstrable might of Korean industry on the international stage, it is still the chaebols who lead the way, with an expansive presence in everything from semiconductors to memory chips to batteries. Ikkjin Ahn, the founder of Moloco, another billion-dollar plus tech company that spans the US and Korea, shared with me a contrarian view on the chaebols. For all their faults it's these companies that showed Koreans how to compete with the outside world, and there's still a lot the newer generation can learn from what came before. 'When you are growing up as a kid, you idolize your father,' Ikkjin told me. 'He's the most powerful man. He can do everything. He's a big superman. Then you become a teenager and enter your early twenties, you hate your father. He did this wrong, he did that wrong, he could have done that better. Why is he not helping me? And then when you enter middle age and when you have your own children, you finally understand that they have strengths and they have limitations, but it's not their fault. And finally you understand that you can learn from what they did well, you can begin to value that and realize it's your role to make the better version.'

CHAPTER FIVE

Smart Nation

'If I'm asking myself as someone who's good with computers, what is the most effective pathway for me to make the world better, it would be to try and make governments better and provide technology tools for them to perform better.'

I

In 2005, when Steve Jobs gave his famous commencement speech at Stanford, Xiaodong Li, a twenty-something business student from Tianjin, listened intently from the audience. Li was one of only four students from China admitted to his MBA class at the Graduate School of Business where his peers had found his name so unpronounceable that he had taken to calling himself Forrest, inspired by Tom Hanks's character in *Forrest Gump*, noting later that he identified with him because he was 'not always the smartest person, or the strongest one physically among his peers, but he has a very good heart'. That sunny Sunday morning in Palo Alto, as Jobs told the graduates to stay hungry, stay foolish, Forrest nodded along. As the ceremony came to a close and the crowd dispersed the young international student couldn't shake the feeling that something had shifted inside him. He would rewatch the speech over and over again multiple times a day for months.

One phrase echoed in his mind: 'You can't connect the dots looking forward, you can only connect them looking backwards, so you have to trust that the dots will somehow connect in the future.'

After graduating, Li followed his girlfriend, now wife, who had also studied management at Stanford, to Singapore where she had found a job in finance. Before his MBA, Li had worked as a recruiter for Western companies in Shanghai where he found the conventional career paths taken by most of the candidates whose résumés he was reviewing so boring that he was driven to distraction and turned to gaming in internet cafés late into the night. And so, for his second act, he thought better of the corporate ladder and branched out on his own to launch GG, a gaming company. It bombed spectacularly. The venture had focused on making single-player games just when the industry was moving in the opposite direction, towards more collaborative multi-player formats. The financial crisis in 2008 took out whatever wind was left in the young company's sails and the startup soon went under. Li's first attempt at playing entrepreneur didn't work out, but he left such a strong impression on his investors, who included Toivo Annus, a co-founder at Skype, that they gave him another million dollars for a do over. And so, in 2009, Li, along with two of his friends, also immigrants from China, launched another company, which they called Garena, a mashup of global and arena.

Garena swiftly went through three incarnations. It first came up as an online gaming platform with a chat feature, sort of like a social network for gamers, where players could play and talk smack at each other, and quickly grew to over a million users within its first year. The platform's popularity drew the attention of major game publishers like Riot Games in the US, which partnered with it to distribute its smash hit *League of Legends* franchise in Southeast Asia. Tencent, the world's largest gaming company, soon followed and bought a 40 per cent stake in Li's company giving the fledgling startup financial rocket fuel and instant credibility in the gaming business. And so began the company's second iteration: a distributor of blockbuster games sourced from the rest of the world into Southeast Asia. Garena would go from a gaming platform to game distributor to, in its third incarnation, a bona fide game developer when, in 2017, it launched *Free Fire*. Forrest Li would with his debut title catch lightning in a bottle, producing a global sensation that changed the fortunes of his obscure startup situated at the periphery of not just the gaming world but the periphery of the actual world and eventually go on to alter the landscape of the entire internet economy in Southeast Asia.

Within two years of its launch, *Free Fire* would become the most downloaded mobile game in the world, installed over a billion times, and eventually top the charts of the highest grossing mobile games in the US. At its peak it was played by 150 million people, or roughly 2 per cent of all of humanity, equivalent to the entire population of Russia, every single day. This one game has made more than $4 billion in revenue, almost twice as much as the highest-grossing movie of all time. It did so even though it was a late entrant in the wildly competitive genre of battle royale, the most popular gaming format of the past decade, in which hundreds of players assume the role of fighting characters – assassins, soldiers, hunters, snipers, zombies, all manners of killers – who converge on a map to duke it out in a *Hunger Games*-style contest for survival. *Free Fire* took almost no time to take its place among the top three games in this format, alongside blockbuster hits like *PUBG* and *Fortnite*, much more established titles which between them have made more than $50 billion in revenue – two games that have made more money than the entire GDP of Jordan and Bahrain – and boast more than a billion registered users. *Free Fire* was able to break their dominance by creating a lightweight alternative that deployed similar gameplay mechanics but was optimized for low-end mobile phones which made it a more practical choice for gamers in emerging markets. And so while *PUBG* and *Fortnite* still reign supreme in places like North America and Europe where mobile and internet infrastructure can easily handle the demands of games at the higher end of the performance spectrum, *Free Fire* has cornered the market for battle royale in up-and-coming places like Latin America, the Middle East, Southeast Asia and India.

The torrent of cash released by *Free Fire* bankrolled Forrest Li's ambitions to take on the wider internet economy in his region. In 2015, he branched out into ecommerce with the launch of Shopee, an online everything store that is a crossover of Amazon and eBay in Southeast Asia, a marketplace where people can buy products from each other, like eBay, and from other businesses, like Amazon. Here too the company entered a crowded market late and then quickly established itself as the top player. Shopee is the number one online store in a group of countries popularly known as the ASEAN-6 – Indonesia, Malaysia, Thailand, Vietnam, Philippines and Singapore – and is easily the biggest online marketplace in Southeast Asia, with almost half of all online commerce in the region taking place on its platform. Getting there was no small feat. The ecommerce market in Southeast Asia is the fastest growing in the world with fierce competition among local players having a go at each other, often with the help of overseas backers, in a contest that permits little in the way of rules and scruples, to claim for themselves the spoils of one of the top five ecommerce markets in the world. The Southeast Asian ecommerce wars are sometimes seen as the domestic battle between Chinese tech firms spilling out into the wider region, with Tencent-backed Shopee pitted against Alibaba-backed Lazada and ByteDance-backed TikTok Shop – the same TikTok that took social media by storm has now set its algorithms loose on selling products online. After conquering gaming and ecommerce, Forrest Li, without missing a step, proceeded to waltz into fintech with a third company, SeaMoney, which is a lot like the PayPal of Southeast Asia, the payments backbone for ecommerce in the region.

In 2017, Li brought all three companies – Garena, Shopee and SeaMoney – under one corporate umbrella, Sea, short for Southeast Asia, a reflection of the company's sprawling presence far beyond the tiny island of Singapore where it first arose. The same year, Sea made its debut on the New York Stock Exchange, where it made history as the first Southeast Asian tech company to be listed on the Big Board and the first billion-dollar tech company from Singapore to go public in the US. The IPO would be only the beginning of Sea's ascent and when the global pandemic hit its fortunes took off. With people staying indoors, all three segments in which it operates – gaming, online shopping and digital payments – saw record growth. Investor confidence knew no bounds and by 2021 Sea was the fastest growing company in the world, at one point worth over $200 billion, breaking into the ranks of the 100 most valuable companies in the world. In Singapore it was in a category of one, the most valuable company this tiny island nation had ever produced, worth more than the next three companies listed on the stock market, all of them banks in a country known for its financial sector, combined. Forrest Li went from being a struggling immigrant entrepreneur to the richest person in Singapore, besting a formidable collection of moguls which include Eduardo Saverin, the co-founder of Facebook, who has for long taken up residence on the island, and is in his second act a major investor in Southeast Asia's booming tech scene with his venture fund B Capital. Sea's rising tide also lifted the fortunes of Li's co-founders, both of whom joined the ranks of billionaires, the newest entrants in the club of Crazy Rich Asians.

Back in 2017, on the day that Sea went public at the NYSE, Forrest Li took to the podium to ring the opening bell to mark the company's milestone achievement. Flanked by his senior leadership and cheered on by the crowd that had assembled on the trading floor below, he rang the bell and pumped his fist in the air. Just over his shoulder, projected on the giant LCD screen behind him, was his newly listed company's circular logo with the words 'connecting the dots' emblazoned underneath: three words that make up the firm's official slogan which appear dutifully wherever its corporate emblem is to be found, an echo of that commencement speech at Stanford all those years ago with which Steve Jobs first inspired Xiaodong Li, now Forrest, by way of *Forrest Gump*, to trust that the dots will somehow connect in his future and to always stay hungry, stay foolish.

2

Sea has hit choppy waters since its early tearaway success. As the pandemic ebbed so did the investor optimism that the region was on a path to supercharged digital transformation with Sea leading that charge. The company over-extended itself by entering distant markets and after an expensive but ultimately futile foray into Europe had to withdraw back into familiar territory. It also got caught in geopolitical tensions between India and China and its main money maker, *Free Fire*, was banned in the world's most populous country, its biggest market, as the company was seen as an extension of Tencent and China in the South Asian nation. The company's stock price tumbled, at one point losing over 90 per cent of its value. Sea, after a few calamitous years, has in recent times been on rebound, clawing its way back to being worth over $60 billion and is still by some margin the largest internet company in Southeast Asia. In any other country this company, even in its diminished form, would be seen as a wild success, the sort of opportunity investors dream of coming across even once in a lifetime, but in Singapore it's often still portrayed as a collapsed empire that wanted too much too fast and couldn't keep the train tethered firmly to the tracks. But more than size, it's the speed with which it has gone through the various stages of its abbreviated life that shows the sheer pace of change that has become the norm in parts of Asia. Forrest Li went from failed entrepreneur to launching a gaming platform, which morphed into a game distributor, which then became a game developer, and then an ecommerce giant, and then a payments company, and then a globe-spanning corporate empire and a cultural force which was the main sponsor for the Brazilian national football team and the creator of virtual worlds in which artists like Justin Bieber released their new songs, to almost going bust, and then rise to relevance again, all in the span of fifteen years. It shows how quickly the digital landscape is shifting in Southeast Asia and the outsized role that the tiny island nation of Singapore is playing in effecting that change.

As with any entrepreneurial success story there are two ways of looking at Sea. One is to see it as the story of gritty individuals battling the odds to make big things happen. The other is to see it as a downstream consequence of macro policy choices made by governments. It is possible to draw a direct line between specific measures taken by the Singaporean government, which since the 1980s has been trying to put in place the foundations for a high-tech economy, and this home-grown, globe-spanning tale of operatic proportions that even much larger nations by comparison have struggled to produce. Singapore's decades-long effort to promote high-skilled immigration has been a central element in this strategy. Since the country's founding it has relied on immigrants to augment its domestic talent pool and it's not unusual for even senior civil servants and government ministers to have been born outside the country. Lee Kuan Yew, the country's founder, has written that seven out of nine members of his first cabinet, formed in 1959, were born abroad. In the early years, there were so many foreigners working for the government that one cabinet member told Lee, half in jest, that he was only made prime minister as a concession to the local population, a token hire meant to appease the natives. 'The Singapore pool of talent is finite and limited,' Lee wrote in 1982. 'Singapore has been like the American space shuttle. It has two rockets to boost it into space. We have a powerful Singapore-made rocket. For that extra zip, we had a second rocket, assembled in Singapore, but with imported components. We must try hard to continue to have that second rocket.'

All key players in Sea were foreign talent nurtured by the Singaporean government and owe their presence in the country to this longstanding talent strategy which has spilled over from staffing government roles to building out the private sector. After business school, Forrest Li moved to Singapore, and not China, which would have been a more logical choice for game development, given that three of the top five highest-grossing games ever came from studios owned by Chinese companies, because his wife went to Stanford on a scholarship extended by Temasek Holdings, the Singapore sovereign wealth fund, which in return obligated her to serve a six-year bond with the state-owned company. Li, who was initially hesitant to go, had no choice when his future father-in-law declared 'If you don't go to Singapore too, no negotiation.' Li's two other co-founders, David Chen and Gang Ye, who were also born in China, arrived in Singapore as teenagers straight out of high school under a government-run effort to recruit foreign students through scholarship programmes that has been going on since the 1990s. All four are now Singaporean citizens.

As a small country with no natural resources, Singapore has for its very survival had to rely heavily on improving the quality of its human capital and, when necessary, on importing it from abroad. This is a major plotline in its transformation in one lifetime from one of the poorest nations in the world to what is today one of the wealthiest – the second-richest country in the world in per capita terms after Luxembourg. The country's economic success is often explained as an offshoot of a massive chip on its shoulder: Singapore is unique among nation-states in that it gained independence against its own wishes. This small island was a part of Malaysia until 1965 when the Malaysian parliament voted unanimously to expel it from the federation in what was seen as a racially motivated measure by the country's Malay majority to sever off its only ethnically Chinese dominated region to preserve the racial character of the wider union. And so while in other formerly colonized territories independence was experienced as a moment of rapturous deliverance, a people being born after decades, sometimes centuries of struggle, in Singapore it was the opposite. In a scene who's grainy black and white footage has now become a formative national memory, the news of the country's independence was announced by Lee Kuan Yew, only 42 years old at the time, in a televised press conference during which he broke down in tears declaring it a 'moment of anguish' believing as he did all his life in the union of the two territories. The resulting country, an island half the size of London and populated by less than 2 million people, most of whom lived in slums and squatter settlements, and with no industry or fighting force to speak of, sandwiched between the much larger hostile states of Malaysia and Indonesia, was seen as a non-viable nation-state, likely to be invaded and occupied by a more powerful neighbour, as indeed it had been by the Japanese only two decades earlier.

What happened next is by now well-known. Singapore pulled ahead of its better-heeled neighbours by opening itself up to the world as the gateway to Asia. Foreign companies looking for an entry point into the world's biggest market flocked to the city state which spoke English, had a reliable legal code, stable politics and zero tolerance for corruption – a combination of qualities that is still rare in the region. The country's early enthusiasm to position itself as a reliable partner for global business has paid dividends. It has retained the number one spot on the Economist Intelligence Unit's ease of doing business rankings for fifteen consecutive years, and frequently topped a similar ranking issued by the World Bank which recently ceased publication. According to the Singapore's Economic Development Board this favourable regulatory environment has brought 59 per cent of the world's tech multinationals and 88 of the world's 100 top tech companies to set up operations in their territory, which include Amazon, Google, Microsoft, Meta, Alibaba, Tencent, ByteDance and hundreds of other tech headliners. The talent and know-how seeping out of these companies has seeded a vibrant startup ecosystem around them. Many of these companies have converged on the One-North neighbourhood, a reference to the country's location one degree north of the equator, a district that is home to Sea Ltd and Google's Asia Pacific headquarters. Block71 is also here, a factory building that has become the heart of Singapore's startup scene, making it to the city-state a lot like what Station F is to Paris. *The Economist* once called it 'the world's most tightly packed entrepreneurial ecosystem, and a perfect place to study the lengths to which a government can go to support startup colonies'. One-North has become one of the trendiest tech districts in all of Asia, with frequent comparisons made to King's Cross and Kendall Square. Designed by Zaha Hadid Architects, the precinct is dotted with gleaming glass and curved metal structures in zones with sci-fi-inspired names like Biopolis, Mediapolis and Fusionopolis, where cafés have robotic baristas and the environment is purpose-built to blur the line between work, life and play.

Singapore took the developmental model of central planning directed by a strong government, the norm in the post-colonial world, further than most with an almost all-encompassing presence of the state in nearly every facet of daily life, a mode of governance that has often been characterized as soft dictatorship or authoritarianism lite. This bureaucratic control extends deep into the lives of its citizens with regulations governing everything from gum chewing and bird feeding to flushing public toilets which remain a part of the country's legal code and are actively enforced even today. The state has a say in everything from where a person lives to who lives next to them by imposing strict ethnic quotas to prevent the formation of racial enclaves and designing housing that forces social interaction among ethnically diverse neighbours. Lee was unapologetic about these apparently illiberal traits, maintaining that an expansive state presence was necessary to create a 'first world oasis in a third world region', noting that: 'I am often accused of interfering in the private lives of citizens. Yes, if I did not, had I not done that, we wouldn't be here today. And I say without the slightest remorse, that we wouldn't be here, we would not have made economic progress, if we had not intervened in very personal matters – who your neighbour is, how you live, the noise you make, how you spit, or what language you use. We decide what is right. Never mind what the people think.'

This dirigiste tendency and all-encompassing role of the state in public affairs has carried over into the country's approach to building a tech economy. In most countries, the startup scene has come up at a distance from or sometimes despite the actions of state authorities. In Singapore, it is the government that has been leading the charge to make the institutional and cultural changes necessary to make the environment more conducive to entrepreneurship, sometimes in the face of a reluctant population which still has a strong preference for stable jobs in traditional careers. The government practically built the venture capital industry from scratch in 1999 with the formation of the Technopreneurship Innovation Fund, or TIF, setting aside a billion dollars to invest in major overseas venture capital funds as a limited partner in a bid to attract them to establish operations in this small country that was at the time peripheral to their concerns. Exposure to the inner workings of top-tier funds allowed the government to rapidly build up its own expertise in venture investing, with Singapore government-linked venture funds eventually becoming early investors in regional giants like Alibaba and Baidu. This early start has made Singapore the clear leader in the Southeast Asian venture landscape. The country is second only to Brunei as the smallest of the eleven countries that make up the Southeast Asian region – compare the 5 million or so people in Singapore to the 270 million in Indonesia or 115 million in the Philippines – and yet it has more private equity and venture capital assets under management than all the others combined.

Much of this push to have more startups is prompted by economic necessity. Singapore's growth has been impressive but also unusual in that it has been driven largely by government-linked corporations or foreign multinationals with a relatively underdeveloped domestic private sector outside of finance. Few would be able to name a prominent Singaporean company that's not a bank or some other kind of financial institution, an attribute which differentiates it from other stories of breakout national success in Asia like China and Korea whose rise was enabled in large part by their vibrant domestic private sectors and whose companies, especially in the high-tech sector, are known the world over. Singapore is richer than both but has not produced much in the way of a Tencent or Samsung, its only noteworthy exception being Creative Technologies, the now lapsed manufacturer of speakers and sound systems, a name that would be familiar to anyone who had a PC in the 1990s. And while the country is generally perceived to be a technologically advanced nation, cue the billion-dollar glass and steel Rain Vortex at Changi airport, its domestic companies don't produce much in the way of scientific publications or patent filings, two of the most basic metrics used to measure innovative capacity, with most of the output in these measures being driven by public universities and foreign multinationals. 'This skewness underscores the low progress made by Singapore toward developing the R&D capacity of its indigenous firms,' notes a recent assessment of the country's innovation and entrepreneurship ecosystem by the World Intellectual Property Organization.

It is this discrepancy that the government is trying to address with a slate of initiatives to produce the sort of fast-growing tech startups that have been driving tech competition between the US and China. Importing the best talent is just one element of the strategy. Throwing money at the problem is another. There's an entire alphabet soup of government grants for Singaporean entrepreneurs that have made the country among the better funding environments for early stage ventures anywhere in the world. Ease of access to funding was cited as one of the key reasons by Startup Genome for placing it seventh in their ranking of best places to start a company. The government has three main financial schemes for companies at various stages of their maturity. The Startup SG Founder provides $5 for every $1 invested by first-time founders to get their company off the ground, a grant which, from what I've heard, is not so hard to get as long as one is affiliated with a Singaporean university. More mature companies that can show a demo can tap the SG Tech initiative which offers up to $380,000 in grants. And for market-ready companies that can raise private capital, the government co-invests at a ratio of 7:3 up to $6 million for deep tech companies. And then there is a slew of other incentives such as tax exemptions on the first $75,000 made by the startup, which is on top of the 17 per cent corporate tax rate which is already among the lowest in the world.

These measures have made the Singapore government one of the most generous benefactors of tech entrepreneurs anywhere in the world. In fact, while in other countries the most frequent complaint is that there is too little government financing for new companies, in Singapore the criticism is that there is too much, creating an entire cottage industry of grantrepreneurs, founders who go from grant to grant, who are entirely reliant on government support for their survival. According to a report by the World Bank, 69 per cent of Singapore-based startups benefit from one government scheme or the other. 'Many companies that should be closing down may in fact be receiving undeserved support from the government,' it notes. A 2017 report by the National University of Singapore found that Singapore-based startups have a better survival rate compared to their counterparts in the US and the UK, which to the untrained eye might seem like an indication that their domestic entrepreneurs are playing the startup game better than those overseas, but on closer inspection is likely a sign that bad ideas are being kept alive longer than necessary. This is prolonging time to failure which is slowing down the rate at which entrepreneurs can learn from their mistakes and re-enter the market with newer and better ideas. Excessive government funding also inflates valuations, which can have the unintended consequence of discouraging private investment.

The Singaporean government's largesse often extends beyond its borders to companies in the wider Southeast Asian region. This again is prompted by necessity. Even if Singapore were to go all out on backing every single domestic entrepreneur to the maximum, the city-state is just too small to make much of a splash on a global level, and so it must turn itself into a hub for regional companies with global ambitions. If in the past Singapore became economically relevant by serving as an outpost for Western multinationals wanting to expand into Asia, the next iteration of its growth relies on its attractiveness to Asian firms looking to conquer markets abroad. The country's involvement in Grab shines a light on its future role in building out the region's tech economy.

Grab was the brainchild of Anthony Tan and Tan Hooi Ling while they were doing their MBAs at Harvard Business School. The company was born when they entered a new venture competition at HBS where they pitched the idea of a ride-hailing service for Malaysia, which is where they were from. Their proposal came in second, with one judge telling them that their home market, a country of 34-million people, was too small for the service to be viable. They went ahead and launched the company anyway, investing the $25,000 prize they had won as runners-up as seed money to launch MyTeksi, which would eventually become Grab, the Uber of Malaysia, in 2012. The company expanded quickly but struggled to tap domestic sources of funding in a highly conservative capital environment. When their application to Khazanah Nasional, Malaysia's sovereign wealth fund, for a $10-million grant failed to materialize, Singapore's sovereign wealth fund, Temasek, swooped in with a $10-million Series A investment, and convinced the founders to move their headquarters to their city-state in 2015.

This would prove to be fortuitous. Grab would see off stiff competition from Uber, which couldn't compete for long given their limited knowledge of local conditions, and eventually the American behemoth would sell its operations to Grab, becoming a partner rather than competitor, with Uber buying a 27.5 per cent stake in Grab and its CEO Dara Khosrowshahi joining the Singapore-based startup's board of directors. Grab would eventually become the biggest technology startup in the region, the first to cross a valuation of $10 billion, and when it went public via SPAC at the Nasdaq in 2021 at a value of $40 billion, it became the biggest such exit in history not just in Southeast Asia but anywhere in the world. For Malaysia, Grab was the one that got away. Had it stayed it would have been the most valuable company the country had produced in any sector ever. Instead, it became one more instance of Singapore upstaging its much larger neighbour and former compatriot, which in living memory had considered it beneath itself and expelled it from the union. Grab has since expanded beyond ride-hailing into food delivery and payments, joining the heated superapp platform wars of the Asian region. The high-profile departure of this one company to Singapore has made this small country a magnet for other companies in Asia looking to relocate to a more startup friendly environment. International VCs interested in investing in places like Malaysia, the Philippines and Indonesia, attracted by their large booming markets but often put off by the chaotic environment, are also more open to writing cheques to founders if they are willing to serve their home markets from a distance, from the more familiar and orderly setting of the city-state.

3

Grab and Sea are similar in that they are the two largest tech companies in Singapore which have grown to the scale of tens of billions of dollars by tapping into the regional market. Grab is the largest ride-hailing service in the region, with a presence in over 700 cities, and Sea is by far the biggest online store in the six countries in the immediate neighbourhood in which it operates. Their ability to serve a growing regional market beyond their home country is crucial to their outlier success. It is here that the future of Singapore's tech sector depends not just on what happens within its own territory but is increasingly tied to the prospects of the wider region. Singapore's domestic economy is highly developed but is and always will be highly constrained by the country's small size. It has little choice but to leverage its position as a springboard into the wider region to amplify its economic and political relevance on the international stage. Nearly all the major startups backed by GIC and Temasek, the country's sovereign wealth fund manager and state-owned investment company, which between them manage over $1.7 trillion, have a focus that in the first instance is best described as regional rather than domestic or even global. Carousell is a mobile classifieds marketplace, Ninja Van is a logistics solutions provider, PropertyGuru is an online marketplace to buy and rent property. All three are based in Singapore. All three are worth over a billion dollars. All three have received state funds from Singapore's two main investment arms. And all three reached scale by primarily serving the dozen or so countries in the immediate neighbourhood.

It might seem obvious and even natural for companies to grow by expanding out into their immediate surroundings except it's something that rarely works out in practice. Geographical contiguity seldom translates into ease of market access, with different countries having their own languages, legal regimes and wildly uneven coverage of quality infrastructure, all compounded by the high likelihood that relationships between neighbours will be fraught with longstanding squabbles. Most of the biggest success stories in tech grew by going global from day one, often skipping the regional and sometimes even their domestic markets. US companies didn't grow by expanding into South America, European companies have had a hard time pretending that the continent is one place, and Indian companies too often skip the regional market to serve primarily Western customers. In this crucial respect, Southeast Asia is different in that its companies often place the region first, an endeavour in which many of them have met with relative success compared to other geographies, and speaks to how a group of countries that are as diverse if not more than those elsewhere and have much in the way of their own geopolitical baggage – Malaysia's fraught history with Singapore being a case in point – can still nevertheless figure out a way to work together to their mutual benefit.

The region in question, Southeast Asia, takes institutional form in the Association of Southeast Asian Nations, or ASEAN, a confederation of ten member countries: Brunei, Cambodia, Indonesia, Laos, Malaysia, Myanmar, the Philippines, Thailand, Vietnam and Singapore. The group has a secretariat based in Jakarta, Indonesia, and is headed by a chair which rotates among member countries every year. The countries have little in common other than geography. The largest, Indonesia, has 280 million people; the smallest, Brunei, has less than half a million. The richest, Singapore, has a per capita GDP of $83,000; while that of Myanmar is around

$1,100. Religious differences are also substantial: from Muslim-majority Indonesia and Malaysia to Buddhist-majority Thailand and Cambodia, and then there are Singapore and Vietnam, among the most religiously diverse countries in the world. Their political systems too range from full democracies to hybrid regimes and authoritarian states. Linguistically it's the most diverse region in the world with over 1,200 different languages being spoken, with Indonesia alone having 800 distinct linguistic groups. And yet, despite being divided along every conceivable fault line, ASEAN, which, with a population of over 660 million, has more people than the EU, has made considerable progress towards economic integration. Intra-ASEAN tariffs are effectively zero and in 2020 the bloc joined Australia, China, Japan, New Zealand and South Korea to create the Regional Comprehensive Economic Partnership (RCEP), the world's largest free trade zone which covers a third of the world's population and produces a third of global economic output. It was seen as a significant milestone towards Asian economic integration which will make it even easier for the region's companies to operate seamlessly across national borders.

China and India receive a disproportionate amount of attention when it comes to the narrative of rising Asia, but in recent years ASEAN too has been quietly gaining relevance. In the past two decades it has been the third fastest growing region in the world after China and India with an annual GDP growth rate of 5 per cent – compare that to 2 per cent in the US and 1.5 per cent in the UK. As a bloc it's the fifth largest economy in the world after the US, China, Japan and Germany. Beyond this secular growth that has been going on for decades, the region's fortunes have also been buoyed by more recent US–China trade tensions. With both US and Chinese companies looking to restructure their supply chains away from the mainland – US to reduce their exposure to China and Chinese firms to skirt US-led export sanctions – ASEAN has emerged as a popular alternative.

The increasing rivalry between the world's two biggest economies is bringing not just companies from these two countries but investments from much of the rest of the world to ASEAN to cash in on China plus one – the widely used shorthand for the newfound necessity for companies across industries to diversify away from the mainland. Singapore invests over $60 billion in Vietnam, with Temasek for instance backing VinFast, a Vietnamese EV manufacturer which already has tens of thousands of its electric cars on the streets. And yet despite this level of investment, Singapore is not Vietnam's biggest foreign investor, that distinction belongs to South Korea which invests even more, with one Korean company alone, Samsung, investing $22 billion in building out its Vietnamese facilities. Indonesia too is developing a mines-to-manufacturing capability for electric vehicles. It's already a major player in the supply chain for lithium-ion batteries with over 40 per cent of the world's supply of nickel coming from this one country. Anyone who is anyone in EVs is invested in Indonesia: Hyundai, LG, BYD, Volkswagen, CATL, all have made billions in investments in Southeast Asia's most populous country.

But the biggest story here is semiconductors. ASEAN is the second-biggest exporter of semiconductors in the world after China with 22.5 per cent of the world's supply coming from this region. Malaysia in particular has been attracting attention. The country has been a major player in the industry for decades; as far back as the 1970s it exported three times as many chips to the US as Taiwan, but eventually lost out to Taiwan and South Korea as companies like TSMC, Samsung and SK Hynix rewrote the rules of the game for the entire industry. But as concerns about overreliance on Taiwan for what is quickly becoming the world's most valuable resource grow, in view of the ever-present danger of armed conflict with the mainland, Malaysia has seen a revival in the fortunes of its semiconductor industry. Penang, an island off the coast of northwestern Malaysia which is a bit bigger than

Singapore, has in particular seen a surge in outside interest. It is one of the few regions in the world that has a long legacy of expertise in semiconductor manufacturing, with its history stretching back to over half a century. An influx of new capital has found its way to Penang with over 350 multinationals now having a presence on this small island which includes three of the top ten semiconductor companies in the world: Intel, Broadcom and Micron. The financial commitments are substantial. Intel alone has invested over $7 billion in building a new chip packaging and testing facility in Penang.

The wider region too is getting attention from the most elite players in the chips industry. In December 2024, Jensen Huang, the CEO of Nvidia, made a high-profile visit to Thailand and Vietnam, announcing a $250-million investment in an AI research and development facility in Hanoi and expressing his intent to make Vietnam a 'second home' for the world's most important AI company. The intent is serious. Nvidia also acquired VinBrain, a five-year-old Hanoi-based AI startup which is a part of the same conglomerate as VinFast. This was Huang's second visit to the region in the span of twelve months. Tim Cook of Apple and Satya Nadella of Microsoft also made high-profile visits to the region on the heels of Huang's visit in 2024, with each announcing a slate of investments, in what was seen as another indication of ASEAN's growing relevance to the mainstream of the global tech industry.

4

ASEAN traces its origins to the Cold War and was initially formed in the 1960s to create a common front against communism on its eastern flank and eventually this security partnership evolved into an economic bloc. This is comforting for Western powers which are courting it again in hopes that it can perform a similar role in the face of a rising China. The Rebalance or Pivot to Asia was a signature foreign policy agenda of President Obama's first term, which signalled a shift in US strategic priorities away from the Middle East and Europe to the Indo-Pacific region, appointing the first American ambassador to ASEAN in 2009 and hosting the first ASEAN summit on US soil in 2016. But in this new era of great power rivalry, the Southeast Asian states have resisted calls to take sides, choosing instead to be ideologically ambidextrous and maximize relations with both parties in what has often been described as a strategy to lean on the US for security and on China for economic prosperity.

If American companies have been increasing their footprint in the region, so have the Chinese, and so far, on balance, it is the latter who have been gaining in relative influence. In 2020, ASEAN pulled ahead of the US to become China's largest trading partner. Ironically it may have been the US that inadvertently midwifed those stronger economic ties. In previous years, Western companies would place orders with firms in Taiwan, who would then ship components to China for assembly, and from there the final products would be shipped to markets in the US and Europe, thereby creating a triangular trading relationship. But with the US now trying to reduce the volume of direct shipments from China, Southeast Asia has been inserted as a buffer into the mix, creating a new 'four-corner' trade model. Components that previously went straight to the US and Europe are now redirected to Southeast Asia to sanitize them from their ties to China before making their way to their final destinations.

According to a recent paper by the IMF, 'Changing Global Linkages: A New Cold War?', the increasing decoupling between the world's two largest economies has led to the rise in the significance of 'connector countries' that are 'stepping in to bridge the gap between rival blocs'. The report raises the question whether severing direct links between the US and China is in fact substantively reducing their mutual exposure to each other's economies if the imports are simply being re-routed through third parties, citing data that shows that reduction in trade flows between the US and China is happening alongside a proportionate increase in investments from China into countries benefiting from improved trade ties with the US. So nothing changes. Except supply chains are lengthened and in effect made more fragile and less secure thereby delivering an outcome contrary to the intent of the sanctions policy. 'The more cross-border flows are re-routed via "connector" countries, the less effective the policies driving fragmentation may be in achieving their stated objectives,' the report notes. There is plenty of real-world evidence that this is happening in practice. Key players in Apple's supply chain like Luxshare Precision Industry Co, AirPods Goertek Inc, and Lens Technology have relocated from the mainland to Vietnam to sidestep 'Made in China' sanctions. Thailand has become a beachhead for Chinese companies trying to ship their EVs overseas.

But it would be misleading to suggest that China's economic ties to the ASEAN have deepened only because of redirected trade flows. The involvement of Chinese companies in the region runs the whole gamut of activities from building out telecoms infrastructure, to selling products, setting up manufacturing facilities, investing in companies, acquiring startups and partnering in joint ventures. Four out of the five most popular smartphone brands in Southeast Asia are Chinese. The top three ecommerce marketplaces in each of the six largest ASEAN economies are either fully Chinese (TikTok Shop), or Chinese owned (Lazada), or founded by Chinese émigrés (Shopee). Six of the eight highest-valued tech companies in the region – Lazada, Grab, Sea, Gojek, Bukalapak, Gojek and Tokopedia – received substantial investments from three Chinese companies: Tencent, Alibaba and Didi Chuxing. Alibaba alone has made major investments of over $100 million in everything from Ninja Van, a logistics company based in Singapore, to DANA, a payments company based in Indonesia, to Mynt, another fintech company based in the Philippines. The US has successfully persuaded countries from Sweden to Japan to restrict the use of Chinese-made equipment in their telecoms infrastructure; but this campaign has encountered notable resistance in the ASEAN with only Vietnam signalling a support for the ban. In 2023, envoys from the US and the EU wrote to the Malaysian government warning that awarding Huawei 5G contracts could compromise the country's national security, in response to which the Malaysian government announced that it would not restrict Chinese involvement.

Singapore has also seen a surge in interest from Chinese firms. Two of the largest tech acquisitions in the country went to Chinese buyers with Lazada, the region's second-most popular ecommerce portal, being snapped up by Alibaba and Bigo Live, a social live streaming platform, being bought by YY. But beyond acquisitions and investments, the form of expansion that has been getting the most amount of attention is the wholesale move of Chinese companies from the mainland to Singapore. The most notable of these arrivals is Shein, which at one point was valued at over $100 billion, and de-registered from Nanjing to move to Singapore in 2021 where its founder and CEO, Chris Xu, is now a permanent resident. On its website the company describes itself as a 'Singapore-headquartered global online fashion and lifestyle retailer' and makes no mention of its origins in Nanjing. Zhang Yiming, the founder of ByteDance, has also moved to Singapore, and the company's most prominent business unit, TikTok, has a Singaporean CEO. These are just the most notable examples among hundreds of Chinese firms which include Nio, the EV manufacturer, and Hillhouse Capital, a major fund known for its early investments in JD.com and Meituan, which have re-domiciled. One motivation is for them to appear less Chinese and thereby skirt heightened scrutiny from Western institutions. The other is to put themselves at a distance from their own government which in recent years has been cracking down on the power of big tech at home.

Singapore too has been under pressure from Western partners to rein in the growth of Chinese companies on its soil. Its main telecoms operators, Singtel and Antina, notably chose Ericsson and Nokia instead of Huawei to build its 5G infrastructure, though its ICT Minister, Subramaniam Iswaran, noted in an interview that they 'never explicitly excluded any vendor'. It is an open question whether Singapore can definitively downgrade economic ties with China to the extent that is sometimes expected of it. The relationship between the two countries is more complex than that of most places. After all, 75 per cent of the island's population is ethnically Chinese. But demographic similarity did not always translate to good ties. Singapore was among the last countries to officially recognize the legitimacy of the People's Republic of China, holding off on establishing diplomatic relations until the 1990s. Lee Kuan Yew actively resisted efforts to characterize his country as the 'Third China' and on his first visit to the mainland in 1976 insisted on conducting official business in English even though he was conversant in Mandarin. Conversely, Chinese media often referred to Lee as a 'running dog of imperialism' for his staunch support of the US during the Cold War, a charge that was often made well into the new millennium. Bilateral ties would get warmer over time and Lee would become one of the few world leaders to have met with every Chinese leader of the modern era, from Mao Zedong to Xi Jinping. He had a particular fondness for Deng Xiaoping who he often called the most impressive statesman he had ever met and thought that without him PRC would have met the same fate as the former Soviet Union. After initially holding his country's Chinese heritage at arm's length, Lee would in later years embrace it more fully, saying that China and Singapore enjoy 'a very special relationship' and noting that 'we are different like the New Zealanders and the Australians are different from the British'.

If Chinese companies have benefited from their access to Singapore, Singaporean firms too have gained from their involvement in China. Temasek and GIC have invested significant sums in dozens of Chinese companies which include Alibaba, Tencent and BYD. According to Singapore's Ministry of Foreign Affairs, China has become Singapore's largest trading partner and Singapore in turn is China's largest foreign investor. In 2024, during a visit by Singapore's Prime Minister Lee Hsien Loong to Beijing, the two countries announced that they were upgrading their ties from 'All-Round Cooperative Partnership Progressing with the Times' to the decidedly more vigorous sounding 'All-Round High-Quality Future-Oriented Partnership'.

5

Singapore has been making headway in building an environment more conducive to startups and even more so in making itself a more attractive destination for overseas tech companies. But when it comes to visible demonstrations of best-in-class examples of how new technologies are deployed in the world, it distinguishes itself primarily not by way of its private companies but through its public sector. The city-state is often seen as something of a model for how new technologies can be used by governments to improve the efficiency of their internal operations and in the delivery of public services. It frequently ranks at or near the top of IMD's Smart Cities Index which measures how effectively cities integrate technology into urban infrastructure.

In most countries, the use of new technologies by public authorities is often limited to the digitization of routine administrative processes, sometimes called e-government. What this usually means in practice is that people can do things like submitting essential forms online and pay for public services like transport and utility bills with mobile payments. These measures are usually seen as piecemeal rather than a step change in how government operates. A thin digital layer is bolted on top of legacy processes which, under the hood, remain analogue but there is no fundamental from the ground up reimagining of how the work that needs to be done could be done better – or, preferably, made obsolete altogether – with new technologies. Furthermore, these isolated projects are usually scattered across many different government functions, an app here, a process automation system there, but the whole thing doesn't necessarily come together in one cohesive, centrally coordinated system-wide programme.

Singapore, where commerce is taken seriously, but still undoubtedly plays a supporting role in what has long been the city-state's central concern, to explore the outer edge of what it means to have effective public administration in a city that always seems two steps ahead of others in defining what a city ought to be, a kind of New Atlantis made real, takes a more comprehensive whole-of-government approach to the adoption of new technologies. In keeping with its legacy of the developmental state model of actively engineering social change by unapologetically reaching deep into its citizens' lives, the benevolent big brother is not leaving things up to chance when it comes to upgrading the technological substrate on top of which its society is built, considering as it does technological adoption to be as fundamental a concern as the economy or governance, and hence a legitimate and necessary avenue for direct government action. This means everything from having sensors in streets to monitor and direct the flow of traffic in real time, to using drones to survey areas affected by the outbreak of disease, to conducting almost all its interactions with its citizens online instead of in physical locations, and deploying motion sensors inside public housing to monitor the health and well-being of the elderly.

The framework for this strategy is the Smart Nation initiative, launched in 2014 to 'build better, meaningful, and fulfilled lives for Singaporeans through technology'. The plan, which was renewed in 2024, lays out an ambitious strategy for the digital transformation of the country along three axes: digital government, digital economy and digital society. It's the most recent of a series of government-led initiatives which began in the 1980s with Singapore's first National Computerization Program which was a push to figure out ways to rapidly adopt new technologies in the functioning of the government. The initiative is coordinated centrally, directly under the Prime Minister's Office, under an umbrella institution known as the Smart Nation and Digital Government Group, or SNDGG, which contains the country's principal digital capabilities, the two main subdivisions of which are the Smart Nation and Digital Government Office, or SNDGO, which operates like the planning wing, and the Government Technology Agency, or GovTech, which is the main implementation arm, a technically capable unit of the government which attracts the best of brightest technical talent of the country, sort of like a Google within the Singaporean government.

The parallel is not a superficial one. GovTech operates very differently from a traditional government outfit; its culture is more like that of a tech startup and its office looks the part, foosball tables, giant LCD screens and the like. In building out GovTech to serve its deep-tech engineering needs, Singapore is different from most governments. While others typically outsource these functions to outside contractors, Singapore has had a firm policy on developing these capabilities in-house and the agency has over a thousand developers working for it. When it comes to evaluating the most exciting things that are happening in technology in Singapore, it is as relevant to look at what's happening in GovTech as it is to look at startups or big multinationals that are based there.

GovTech has taken the lead in rolling out Strategic National Projects, or SNPs, that form the backbone of the Smart Nation initiative, the most notable of which is Singpass, the country's national digital identity system, a platform through which citizens can access over 2,000 services from over 700 government and private sector organizations. Ninety-four per cent of all government services can be accessed through the app which includes functions like filing taxes, applying for a passport, getting a driving licence, booking medical appointments, applying for a loan, and even public housing. Over 4.5 million people, or 97 per cent of the eligible population, uses Singpass, making it one of the most widely adopted digital identity systems in the world. Other SNPs include the Smart Nation Sensor Platform, a nationwide network of sensors which brings the Internet of Things to urban services and monitors everything from water leakage in pipes, to air quality, to rainfall, and even whether someone might be drowning in the city's many public pools.

GovTech also operates data.gov.sg, an online platform through which it makes over 5,600 datasets from 71 government agencies, like environmental monitoring data and real-time traffic information and real estate market analysis, freely available to the public which they can use to build their own services on top of, like for instance creating location-based services using the government's geospatial data. When not leading initiatives itself, GovTech works with other agencies and private companies, among the biggest of which is Virtual Singapore, a 3D digital model of the city-state, the first digital twin of an entire country. The 3D model, which uses real-time topographical data, can be used for planning purposes, like simulating how building roads and bridges will impact traffic flow, or model the construction of new neighbourhoods virtually to see how they might impact the environment before they are built. Even regular people can use it, like for instance home buyers who want to check how many hours of sunlight an apartment gets before going ahead with their purchasing decision.

Singapore is constantly learning from other countries to adopt the best ideas in play elsewhere in the world into its own operations. I went to Singapore some years ago on the invitation of GovTech to participate in an event called Digital Government Exchange, or DGX, a forum for countries that are on the leading edge of adopting technology in government to come together to share experiences. One morning they took us to the Tuas Port, a $20-billion megaproject which opened in 2022 and when it is fully completed will be the world's largest fully automated port. At the time of our visit, in the sections that had been completed, there was hardly a worker visible on site, with loading and unloading of shipping containers taking place with automated guided vehicles (AGVs) and even ground transportation being done by unmanned driverless vehicles, all managed remotely from a control centre. In economic terms this is one of the most critical projects currently in progress in Singapore given the outsized role the country plays in the international maritime industry. Singapore is the world's second busiest port in terms of total shipping tonnage, up to a fifth of all containers in the world pass through its straits, and half of the world's annual supply of oil.

6

Singapore is known for channelling its best talent into the public sector and with GovTech it's no different. The government aims to staff the agency with the same calibre of talent that would make its way to Facebook or Google, even going to the extent of trying to poach talent from these firms to staff this unit. Lee Hsien Loong, the country's former prime minister, would make frequent visits to Silicon Valley where he would make it a point to reach out specifically to Singaporean talent working at big tech firms to try to get them to come back and work for government outfits like GovTech. Those who couldn't be immediately convinced to make the jump were encouraged to try out working for the government for shorter stints, like with the Smart Nation Fellowship, a three-month to one-year programme to get talented Singaporeans to come back and spend some time with public agencies like GovTech. The fellowship is a useful on-ramp to get people involved. Some do their tour of duty and go back, others decide to stay, and both are desirable outcomes for the government, since they both bring fresh new ideas to the working of state institutions.

The appreciation for technical talent runs deep in the Singaporean government, even in managerial positions, which seems like an unremarkable thing to say, except it's not. In most public sector institutions the world over, staffing skews towards people from economics and legal backgrounds rather than those who come from science and engineering. In 2015, the then Prime Minister, Lee Hsien Loong, who has a first-class degree in mathematics and a diploma in computer science from Cambridge, shared a Sudoku solver on Facebook which he had programmed himself in C, which racked up 36,000 likes and 12,000 shares within a day. One of his ministers, Vivian Balakrishnan, promptly translated the program to JavaScript and posted his own version online. That sort of alacrity for technology in the upper echelons of government is still rare in much of the rest of the world and Singapore has that to its advantage.

The system is under pressure though. The government may have assembled an enviable collection of technical talent in public institutions, but with the growing ranks of global tech companies opening offices in Singapore – like OpenAI did in 2024 – the public sector has to fight hard to keep its people. At DGX, when I asked the head of the Singapore Civil Service, Leo Yip, what was top of his mind when it came to his day to day running of the country's public administration, he said it was how to hold on to his best people. Private companies, especially those from overseas, have a strong preference for hiring people with civil service backgrounds, given that they bring not just technical expertise but also valuable insider connections, and have seemingly bottomless budgets to lure them away from their government jobs. The Singapore government isn't exactly lacking money either, its functionaries are among the best paid civil servants in the world, but there's an upper ceiling for what public servants can pay themselves beyond which things begin to look more than a little unseemly. So the government is in a bit of a bind. It wanted more foreign companies to move there. And now it has that. But now those foreign companies want the government's people. So a government that isn't much used to competing – Singapore being effectively a one-party state with a relatively weak domestic private sector – now has to get creative at figuring out non-monetary incentives to keep its talent in the face of heightened competition for its people from abroad.

'Salary does make a difference but I think it is trivializing and in fact self-defeating to assume that that is the only and most critical factor,' says Hongyi Li, the Director of Open Government Products at GovTech. Li has two degrees from MIT and worked as a product manager at Google before returning to Singapore to work in government. He studied in the US on a government scholarship which obligated him to come back to serve the country, a scheme that has long been an important on-ramp to get more top-tier talent into public service. What started as what he thought would be a brief sojourn in the public sector before he headed back stateside has turned into a more than a decade long career in GovTech. Li tells me that even if he was better compensated in the private sector, those monetary incentives don't quite make up for the sense of purpose he has gained doing what he does now. 'If I'm asking myself as someone who's good with computers what is the most effective pathway for me to make the world better, it would be to try and make governments better and provide technology tools for them to perform better and I would do so by starting with my own government,' he says. 'And I'm not doing this here because Singapore is the one most in need of help. I know there are plenty of countries in the world which are much more in need of help. I'm doing this in Singapore because this is easy mode. For all the frustration and complication of working in the government in Singapore, getting the Singapore government to be efficient is easy mode. And if you can win that and you can get that settled then you have an opportunity to help other people.'

Li, who is still in his thirties, is strongly invested in Singapore given his grandfather, Lee Kuan Yew, founded the country, and his father, Lee Hsien Loong, served as prime minister for twenty years. Singapore's place in the world looms large on his mind and he thinks that the country's hard-earned prosperity would be hard to sustain without better talent in government. 'I find it offensive that as a society we are allocating our most precious resource, which is brainpower, in the stupidest way,' he tells me. 'The way I phrase it is that the smartest people are working on the dumbest problems.' It's a realization he arrived at while working at Google, where even trivial details were poured over with the most intensive engineering methods, but none of that sophistication was being brought to bear on much more consequential problems like health and education policy. 'There was a shift in my perspective,' he says. 'The shock I would say came when I saw that at Google if you are trying to figure out the shade of blue for the ads there are literal teams of PhDs that run all kinds of experiments to control for all kinds of stochasticity, all the stuff you learned in school about controlling for all this heterogeneity and variables, all these things that you do because every mistake you make costs hundreds of millions of dollars. And you're full of capability and you feel very empowered to solve really small things. But the way I think about it is, let's say you were amazingly successful, let's say you somehow managed to double web traffic and double their ad revenue, you would be some kind of messiah in the Valley, but if you made a national announcement about that, that people are searching Google twice as much, clicking on ads more, people will be like, who the hell cares? I think that put it in context for me, which is that even this absurd success that you could have of getting people to search the web twice as much and click on twice as many ads would just be nothing in the context of actual society despite the money it makes.'

In most countries the government is a bystander or an active impediment to the development of the tech sector, but in Singapore it's the state that's taken the lead in bringing more technology to itself and the wider country. So why can't that model of public sector-led innovation, or, on a more basic level, just having more effective public administration in general, be adopted elsewhere? Why can't we just take what Singapore is doing and do it in more places? It's a question that comes up often, almost as often as the question of why more countries can't have something like a Silicon Valley: shouldn't it be about just doing what they're doing over there and do it over here? And the answer in both cases is the same: it's hard because they operate in a fairly unique environment. The Singapore example is not that easily replicable because it has the benefit of having a single layer of government in the entire country, the city is the state, which makes things a whole lot more manageable. The very fact that it can even conceive of an all-encompassing technology strategy in the form of Smart Nation is because it doesn't have to contend with a stacked bureaucracy that goes from the federal to the state, district and municipal levels which would make policy coordination that much harder. Singaporean officials would be the first to admit that even they wouldn't know how to replicate the same degree of horizontal and vertical integration they have in their government functions at the scale of a China or India. And, in an observation made to me by

Theo Blackwell, the Chief Digital Officer of London, when it comes to a universally applicable template for smart cities 'Singapore doesn't count because that's cheating' since it can bring national powers to bear on municipal issues. City planners in other countries, even if they decide to put themselves in an administrative parenthesis from the dysfunction of the wider national policy apparatus, still wouldn't have control over things like the currency, fiscal and monetary policy, trade and foreign relations like Singapore does to pull those macro levers to deliver on even the minutiae of its urban planning objectives. So has Singapore done well for itself? Absolutely. But is it a model? Harder to say. More like an inspiration.

CHAPTER SIX

Small Wonder

'Switzerland is a kind of trust, reputation is the brand.'

I

The central paradox of Switzerland would be how to reconcile its lofty perch at the top of just about every innovation ranking out there with the seeming absence of observable demonstrations of this purported ingenuity out in the wild. It reigns supreme in the two-dimensional terrain of x and y axes, a creature of the top-right quadrant, a wild statistical anomaly that soars above all them other nameless faceless data points, that undifferentiated haze of specks, that splotch, a stain, that trails it, in askance, like a swarm of mosquitos or a plume of smoke. And yet in the three-dimensional world of hard facts, self-evident truths and tangible objects that can be kicked and licked, this absolute slayer of the bell curve is curiously nowhere to be found.

Switzerland ranks number one in WIPO's Global Innovation Index, a title it has held not by mere happenstance or fluke once or twice ever so often but for an unbroken chain of fourteen consecutive years, just about as long as anyone's been keeping score, rendering this tiny Alpine nation hands down the apex predator, the undisputed champion, the, if you will, Roger Federer of this entire new geography of innovation enterprise. It's not just the WIPO, the country is also a favourite of league tables put out by other reputable establishments that expend enormous energies on divining what's what with the tyranny of sheer numbers: it tops the charts on the European Commission's Innovation Scoreboard, INSEAD's Global Talent Competitiveness Index, IMD's World Talent Ranking, and also, frequently, the Bloomberg Innovation Index. That's resounding validation.

But this heft, it would appear, can only be counted but not seen or felt. Few examples of Switzerland's much publicized outlier capacity to out-innovate the world come readily to mind. Ask a techie in New York or Shanghai or Tokyo to name a Swiss tech company or technology or even just a popular game or an app or a gadget and they would be hard-pressed to name even one. Watches, cheese and chocolates? Sure. But cool tech? Unlikely. When I asked a London-based venture capitalist, who happens to be Swiss, why he chose to relegate himself from the place that has been anointed, with rigorous statistical analysis no less, as the innovation capital of the world, to instead ply his trade in the relative backwaters of the UK, which, according to the pointy-heads at WIPO, could only muster up enough gumption to place a mere fifth, also ran, pretender who couldn't even manage a podium finish, burning embers and wafting smokey residue of what was once was a practical conflagration of a country, my query was met with blank stares.

It's not just that Switzerland is not much known in tech circles; sometimes it would seem that it's not much known at all, about as incognito as the numbered bank accounts, now long past their heyday, that are still often its calling card abroad. When I asked a San Francisco-based Swiss official, who has broad exposure to the goings-on in the Bay Area and Switzerland, about what Americans most often get wrong about her country, her answer, tinged with barely concealed indignation, was: 'They think we're Sweden!'

Americans are of course acclaimed for their practised and habitual indifference to rudimentary geography: Iran/Iraq, Switzerland/Sweden, Luxembourg/Lichtenstein, Ubeki-beki-beki-beki-stan-stan. Why draw finely calibrated distinctions in this morass of lapping syllables when it is understood, with pursed lips and tight knowing nods, that the only boundary of any consequence is the one that separates USA from not USA. The Swedish tourism board recently commissioned a study to figure out the extent of the problem. It concluded that over half of Americans can't tell the two countries apart. This latent nescience might not rankle so much had it not at times erupted into something of a public spectacle.

Spotify was founded in Stockholm, a city which is no less than a thousand miles northeast of the outer reaches of Switzerland (that is up and to the right, for the benefit of my American readers). When it went public at the New York Stock Exchange in 2018 at a valuation of over $25 billion, it broke the record for the largest IPO in the world that year and catapulted itself into the ranks of the top ten tech IPOs ever. The NYSE honoured the breakout Nordic star's formidable achievement by promptly hoisting outside its building not the blue and yellow Swedish flag but the red and white Swiss one instead. Oops. To add insult to injury, the Big Board was more amused than embarrassed by this flub, tweeting later that day: 'We hope everyone enjoyed our momentary ode to our neutral role in the process of price discovery this morning.'

All of which begs the question: Why Switzerland? The US has tech companies valued at multiples of the entire Swiss GDP, its technologies have made their way to every nook and corner of the planet, and, with Voyager 1, the most distant man-made object from Earth, even into interstellar space. China makes more clean energy than the rest of the world combined and with megaprojects like Tengger Desert Solar Park, also known as the 'Great Wall of Solar', which stretches for over a thousand kilometres, has shown itself capable of turning the desert sprawl into one giant power outlet. What has Switzerland done that measures up to all of that? Orson Welles, who plays a racketeer in the movie *The Third Man*, has a line, allegedly borrowed from Graham Greene, that goes: 'In Italy for thirty years under the Borgias, they had warfare, terror, murder, bloodshed. They produced Michelangelo, da Vinci, and the Renaissance. In Switzerland, they had brotherly love, five hundred years of democracy and peace, and what did they produce? The cuckoo clock.'

Harsh. But maybe not unfair? Switzerland does have a not entirely unearned reputation for being a bit . . . sterile. Too rich, too clean, too stable, and too officious in regulating trivialities like, for instance, the rules governing the flushing of toilets after 10 p.m., to conjure the restless impulses that can will vast new forms to life; typifying attributes that add up not to a frothy witch's brew of invention but, it would seem, something that approximates to its exact opposite.

2

It's not just Switzerland's place in the innovation rankings that is counterintuitive, it would seem its very existence as a separate area on the map demands explanation. How is this place even a country? For starters, its three dominant ethnic groups – Germans, French and Italians – have a lot more in common with their respective neighbours Germany, France, and Italy than they do with each other. And yet instead of seeking union with their larger cultural groups, which would make eminent sense, they come together as a separate country with a strong sense of national identity. It's one of the smallest countries in the world, only 8 million people, the same as New York City, and yet its inhabitants speak four completely different languages. And despite its small size, the whole country is about as big as the San Francisco Bay Area, its twenty-six states, called cantons, are highly autonomous with their own constitutions, governments, parliaments, courts and police forces. Imagine Silicon Valley divided into over two dozen administrative zones effectively acting like their own little statelets; Palo Alto speaks a different language from San Francisco, Mountain View has its own parliament which is separate from the one in Cupertino. In just about every country politics converges on two, maybe three, major parties. Switzerland has eleven of them in parliament. But perhaps its most distinctive feature is that it doesn't even have a head of state like, you know, normal countries do, and supreme political authority rests with a council of seven people, called the Federal Council, where decisions are made by consensus. It's all very *Lord of the Rings*. And yet despite this fragmentation the system works remarkably well. Switzerland is, in per capita terms, among the top five richest countries globally. It has more Nobel Prizes in the sciences (23) and more Fortune 500 companies (11) relative to population than any other country in the world. And so, while in the Orson Welles caricature, a perception shared by many, Switzerland is this sanctuary of radical conformity, a place where even minor deviations from the norm like, say, disposing a blue glass bottle in the green glass bottle recycling bin, would constitute a scarcely conceivable subversion of public order, the reality is that the very idea of Switzerland is a contrarian one, a fundamentally different approach to ordering society so unique that it stands out globally. It has had the confidence to take an approach to governance that would likely fail in most countries and made it work; crafting a system which takes what ought to be its biggest weakness, a fragmented polity, and turning it into a formidable strength.

What does that have to do with innovation? A lot. While the rest of the world copies what works from each other – you have a two-party system, I have a two-party system, you have partisan competition, I have partisan competition – Switzerland breaks from consensus, it does things its own way, and in that it is an original from the ground up. On the most fundamental questions of how to design institutions that govern the workings of a well-functioning society, it is different, it innovates. And this nonconformist streak also animates its approach to building out new technologies and the systems that produce them. Switzerland might not have the usual trappings of what we have typically come to equate with innovation, billion-dollar funding rounds and zillion dollar companies, but maybe that's a reductive way of looking at things to begin with. It's not a cop-out answer; it taps into important questions that we asked in the beginning. Can innovation look different in different places? And does it need to evolve beyond what we currently understand it to mean? And it is here that the Swiss example makes for such a compelling study; it's this little petri-dish of other possibilities. It's not that Switzerland hasn't built big, bold monuments to human ingenuity, it has, they just happen not to look like Apple and Google. It's not solely the innovation of companies and products, though they have plenty of those too, and maybe that's a good thing. Switzerland's technological acumen is best exemplified not by the cuckoo clock, whose charms, such as they are, are not in fact Swiss, but German, but by large-scale systems like its cutting-edge transport network, scientific institutes like CERN, and the way it organizes its universities: three domains – transport, science, and education – in which it punches at the highest level globally while adopting an approach that is entirely its own. When it comes to breaking the mould, the Swiss do it their own way.

3

Switzerland has the densest rail network in the world, about 200 miles of train tracks for every 1,000 miles of territory. It's not just big, it's state of the art. All Swiss trains, 100 per cent, are fully electric and have been that way since 1967. What's more, 90 per cent of that electricity comes from renewable sources, and the entire network is expected to run fully on clean energy within this decade. In this respect, the Swiss practically live in the future. Even today only around 1 per cent of the trains in the US, a third in the UK, and just over half in Europe are electric powered. EV manufacturers won't let us forget that they're not just selling a product, but the promise that we can wean the world off fossil fuels. But the environmental benefits of weaning the world off car-centric transportation networks are even more pronounced. Which would make electric trains one of the most effective climate technologies available today. The fact that the Swiss have been able to deploy them at scale ahead of just about any other nation speaks to not just their ability to develop groundbreaking new tech but also their receptivity to technological change. Which is no small thing. Just because something is better and it works doesn't make it inevitable that it's also going to be widely used. Techno-scepticism, cultural resistance to change and the entrenched power of corporate interests invested in preserving the old order have hindered many a technology from widespread adoption.

The Swiss have been building out their rail network for over 175 years because for them it's not just a mode of transport but the infrastructure backbone that holds the whole place together. Being ahead of the curve is not a preference, it's a necessity. Switzerland has one of the most challenging topographies of any country – the Jura on top, the Alps on the bottom, and a central plateau in the middle, which is where most of its people live. Train networks weave together the remote towns and villages of this small but uneven country into a coherent national unit. The need to build transport links through some of the most demanding geological conditions on earth have also meant that the country has had to develop strong engineering capabilities. These are probably best exemplified by the Gotthard Base Tunnel, which connects Zurich to Milan, for which they had to drill a 57-kilometre hole, two kilometres under the Alps, the world's largest and deepest traffic tunnel, which took over $12 billion and seventeen years to complete. The Gotthard Base Tunnel is considered one of the largest and most sophisticated infrastructure projects in modern history, an achievement comparable to the Panama Canal or the Three Gorges Dam in China. And this is why even if trains are a subject that doesn't inspire much excitement in most other places, in Switzerland they are considered central to the country's national identity, a symbol of what they consider quintessentially Swiss ideals of engineering excellence, environmental consciousness and democratic access. Not just a way of going from point A to point B but a means of unifying a dispersed society and defining its culture.

There are other benefits. Switzerland relies heavily on trains because they don't just reduce traffic and pollution but also help contain urban sprawl and rents. They take pressure off cities by drawing the residents away from crowded city centres and spreading them out further into the suburbs and countryside. The population of Geneva, the country's second largest city, is five times higher during the day than it is at night, with commuters coming in from outlying regions, even neighbouring countries like France, who shuffle into town in the mornings and then head out in the evening. This dispersion of population away from the cities, enabled by a dense train network, has meant that there are no large cities in Switzerland. Its largest city, Zurich, a major global financial centre, has less than half a million residents, which is fewer than Omaha, Nebraska. The centrality of rails to life in Switzerland have made the Swiss the most intensive users of trains compared to any other people on earth, with an average person taking fifty-three trips covering over 2,000 kilometres every year. A third of the entire population lives within 5 kilometres of just one train line, the rail system's central artery which connects the two ends of the country, from French-speaking Geneva to German-speaking St Gallen.

The Swiss train system is not exactly Google but it's a vivid demonstration of how cutting-edge tech can take different forms in different places. It doesn't necessarily have to be private and commercial; it can also be public and civic-minded. The historian Tony Judt has written that 'to travel in Switzerland is to understand the ways in which efficiency and tradition can seamlessly blend to social advantage'. This is reflected in the unique corporate structure of the Swiss National Railways. It is a publicly listed company, but all shares are owned by the state. Which goes some way to explaining how it can deliver private sector efficiency at public sector prices. Ninety-two-point-five per cent of Swiss trains are on time, the most punctual in Europe, and the starting point from which this small Alpine nation derives its wider reputation for technical precision and reliability.

4

Sceptics could perhaps be excused for not getting as excited about trains as the Swiss. Is there anything more inspiring going on here? The answer to that is also yes. Buried a hundred metres under Geneva is the tunnel of the Large Hadron Collider the giant ring of which loops across the border between France and Switzerland crossing this international boundary six times along its 27-kilometre circumference. The LHC is the largest machine made by humanity and as a technical achievement ranks among the most complex pieces of equipment ever built, right up there with the International Space Station and the Hubble Telescope. In this tube-like structure, subatomic particles are accelerated close to the speed of light using superconducting magnets chilled to a temperature colder than outer space and made to collide with a precision which is like firing two needles 10 kilometres apart in a way that they meet head-on halfway. These collisions can reproduce conditions that existed within a billionth of a second of the birth of the universe and have helped scientists gain a better understanding of fundamental physics.

The Large Hadron Collider is obviously an impressive machine. But the institution that built it is an equally remarkable creation. CERN, the European Organization for Nuclear Research, is one of the world's most respected research institutions and something of a gold standard for what scientific collaboration across borders could look like. At a time when the world is fragmenting further in research and development with fears that this nation or that could take the lead in fundamental technologies, CERN achieves the unlikely feat of bringing countries together, some of whom are bitter rivals, to pool resources and talent in pursuit of big science. When I asked Yoshua Bengio, the world's most-cited computer scientist, if their grand project of building Pan-Canadian AI institutes has a whiff of CERN about them, his answer was a wistful 'I wish!'

The significance of CERN and other institutions like it goes beyond science. They have often also been the incubators of new technologies. In the US, the military has played an outsized role in building out fundamental technologies which later find their way to civilian applications: the internet, GPS, microwaves and radar being the most well-known examples. In Europe, the military has made fewer major contributions to breakthrough technologies. For starters, they're just not as well-funded. DARPA has a budget of over $4 billion. Its European equivalent, the Joint European Disruptive Initiative, or JEDI, tops out at €100 million. But it's not just about the money. Europe has long had an ambivalent outlook towards military spending. Its inclination is to favour big science over military research with the hope that institutes like CERN can incubate new technologies just as effectively as large military bureaucracies, an approach that would be in keeping with the region's more pacifist temperament.

It's not just CERN. Europe has also been funding other big science projects with the hope that technologies spun off from them can help the continent's competitiveness. The International

Thermonuclear Experimental Reactor, or ITER, the world's largest nuclear fusion research project, is currently under construction in the south of France at a cost of over €20 billion. The billion-euro Blue Brain and Human Brain Projects, designed to improve the understanding of the brain, which, like CERN, were also based in Switzerland, had similar ambitions in mind.

These are not just abstract scenarios. Research from CERN has made its way into applications as diverse as magnet technology, cryogenics, cancer therapies and medical imaging. But the big one is the World Wide Web. It was while working at CERN in 1989 that the British scientist Tim Berners-Lee invented the web, initially conceived as a way for scientists at CERN and elsewhere to share information more easily. It was in Geneva that the world's first website went live, info.cern.ch, on Berners-Lee's NeXT computer. The revolution that started in Geneva has now gone global with over half of all humanity connected to the web. The fact that the world's first website was not a dot-com or a dot-net or even a dot-gov but a dot-ch, which stands for *Confederatio Helvetica*, Switzerland's Latin name, tells you all you need to know about the outsized but often discreet role this small Alpine nation plays in bringing forth consequential new technologies.

Like the railways, here too there is a civic-mindedness at play. In 1992, when there were still less than fifty web servers in the world, Berners-Lee was faced with the choice of patenting his invention for CERN, which would have been the norm, or to leave and make a commercial company out of it. There were even discussions around pricing for such a service. But he eventually decided in favour of making it open and freely available, with CERN relinquishing its intellectual property rights. This goes a long way to explaining why the web took off in a spectacular way while a competing service, Gopher, which had initially been far more popular and technically superior to the World Wide Web, but which charged a licensing fee, stagnated. Tim Berners-Lee would later remark, 'had the technology been proprietary, and in my total control, it would probably not have taken off. You can't propose that something be a universal space and at the same time keep control of it'. When he was honoured at the 2012 London Olympics he appeared on stage and tweeted, 'This is for everyone.'

5

Particle accelerators seem like an exotic hard-to-fathom technology but in fact there are quite a few of them out there, by one estimate over 30,000, with the Large Hadron Collider being the largest of them all. This relatively obscure device is not, however, impervious to international competition. CERN is planning an even larger one, the Future Circular Collider, or FCC, also based in Geneva, which, at a circumference of 91 kilometres, would be three times larger than the LHC and ten times more powerful. Not to be outdone, China is planning an even larger 100-kilometre Circular Electron Positron Collider (CEPC), which, if built, would be the largest in the world.

CERN is of course an international effort involving over two dozen countries. Switzerland has another collection of particle accelerators at the Paul Scherrer Institute, or PSI, in Aargau which is an entirely Swiss undertaking. Christian Rüegg, the institute's director, uses these devices to shoot high-energy particles at materials to study their quantum properties. The institute also runs a cancer treatment facility where proton beams are directed at deep-seated tumours to destroy them while sparing surrounding tissue, a gentler alternative to more aggressive treatment options. PSI maintains other large-scale machines for scientific experiments which have been used to develop future technologies like instruments used by space agencies in the US, Europe and Russia to detect radiation in space and radiocarbon methods to date archaeological discoveries.

PSI, which has an annual budget of half a billion dollars, is just one example of the importance that Switzerland places on scientific research. This small country has the highest number of scientific publications per capita, the highest number of patents filed per capita, and, barring small statistical anomalies like St Lucia, the highest number of Nobel Prizes per capita of any country in the world. It's also among the top five nations who spend the most on research and development as a percentage of GDP.

But this enthusiasm for research hasn't always extended to a proficiency at reaping its commercial dividends. Some of that is just cultural. The Swiss scientific establishment, conservative by temperament, has traditionally held itself to a higher standard than most other places when it comes to conflict-of-interest concerns, preferring to maintain a healthy distance from profit-making schemes. 'It was not built into the mindset of people,' Rüegg tells me. But that is slowly changing. A new innovation park is coming up next to the PSI campus, one of six in Switzerland, where entrepreneurs can set up shop to bring PSI's research into real world applications. It takes inspiration from the Stanford Research Park, built in 1951, which was the initial catalyst for the university's transformation from a research institution to an entrepreneurial school. Rüegg expects that the innovation park won't produce consumer apps and gadgets but be more of a hub for deep-tech startups in material science and engineering. 'The next Uber won't be invented here,' he says.

Switzerland's scientific prowess is built on a higher education system which is rated among the best in the world. Rüegg got his undergraduate and graduate degrees in physics from the Swiss Federal Institute of Technology, also known as ETH, in Zurich, a school which looms large over the scientific and engineering establishment not just in Switzerland but all of Europe. It is widely regarded as the best technical university outside of the US and the UK, a permanent fixture in rankings of the top ten universities in the world and top five in Europe. Alums include Albert Einstein, who did his undergraduate degree here, placing an ignominious second-last in his class, and John von Neumann, who established the mathematical foundation for quantum mechanics, a man about whom it was often said that while most mathematicians prove what they can, von Neumann proves what he wants.

ETH punches at the same level as the top schools in Anglosphere even though it operates very differently in some important respects. The most obvious is the dramatically lower cost of attendance. ETH charges a flat fee of around $1,500 a year for undergraduates. This has made it a magnet for talent not just from Switzerland but from all over the world, and many of these international students stick around to start their own companies. 'I was stoked to go here and pay a fraction of the price I would pay at MIT or Stanford, which I couldn't afford at that time,' Maximilian Boosfeld, originally from Hamburg, Germany, who came to Switzerland for his undergraduate degree at ETH but then stayed to start his own aerospace company, Wingtra, a drone manufacturer based in Zurich, told me. 'It's the best university in Europe, especially in robotics.'

Openness and democratic access, prized Swiss values, are also in evidence in other ways at ETH. In the US and UK, acceptance rates are a key metric for gauging academic prestige. The more students a school can turn away, the better its perceived to be. Admission to ETH at the undergraduate level is essentially uncompetitive. The school is obligated by law to take in anyone with a Swiss high school diploma so the acceptance rate, for domestic students, is effectively 100 per cent. So how does it maintain its academic standing and prestige? In the US, filtering happens before students enter college, in Switzerland it happens after. Top US universities are hard to get into but easy to graduate out of; Swiss schools like ETH are the other way around. Harvard College has a 98 per cent graduation rate. At ETH, over half the students who start don't end up making it to graduation, and in some disciplines, like maths, the dropout rates are even higher.

And so even if ETH is open access, the high failure rate effectively makes it a very selective school. 'If you're a bad student you don't even apply,' says Nathalie Casas, a top official at the Swiss Federal Laboratories for Material Science and Technology, who holds a PhD from ETH in carbon capture technologies. She tells me that out of the 150 students who started in her chemical engineering programme, only a third made it to graduation day.

Is it progress to have a small number of universities that get progressively more selective or is it progress to have a system that can scale a Harvard quality education to everyone who wants to benefit from it? It is often assumed that there is a tension between equity and excellence in higher education, that improving access comes with the inevitable risk of dilution of standards. But as the Swiss experience proves, it is possible to give everyone at least a shot at proving they have what it takes to compete with the best, and that too at minimal cost, without necessarily compromising academic prestige.

6

ETH is an old and well-established institute with its history stretching back over 150 years. In recent years it has been joined in the rankings by its sister school in French-speaking Switzerland, the École Polytechnique Fédérale de Lausanne, or EPFL, which in its current form came up as recently as 1969. It spent three decades in relative obscurity before its fortunes were transformed under the leadership of Patrick Aebischer, a neuroscientist who had built much of his academic reputation in the US while on the faculty at Brown. Aebischer spent sixteen years as president at EPFL, from 2000 to 2016, an unusually long tenure for a university leader in Europe, during which the school elevated its standing from being seen as a keen but middling institution to joining the ranks of the top twenty engineering schools in the world. Known particularly for his fundraising skills, Aebischer was raising over €200 million a year near the end of his tenure, a huge sum by continental standards, because of which EPFL now boasts some of the best scientific research facilities in the world, including a nuclear fusion reactor on its campus, one of only a handful in Europe.

When most universities on the continent still saw themselves as purely academic institutions, Aebischer consciously set out to shape EPFL in the mould of entrepreneurial schools like Stanford. 'It had the Nobel Prizes but it also had the Googles and Genentechs,' he tells me. Venture funding raised by EPFL startups, which was almost nonexistent when he first arrived, stood at half a billion dollars in 2023, second in Switzerland only to ETH. Known for his close ties to industry, Aebischer sits on the board of Nestlé and Logitech and is the chair of the Novartis corporate venture fund. He transitioned out of his role at EPFL in 2016 to focus on startups and investing. He thinks that if an obscure research school in Lausanne can join the ranks of world-class universities in a span of just fifteen years then perhaps the same can happen for the startups ecosystem in the country. 'I'm obsessed about the areas where Europe can be competitive,' Aebischer tells me, 'and if I take one step further, the areas where Switzerland can be competitive.'

But in a world where tech superpowers like the US and China can bring enormous resources to bear on staking out their claim over industries of the future, where can a country like Switzerland, a fraction of their size, with its entire population comparable to that of New York City, carve out a niche for itself? The answer, for Aebischer, is healthcare. 'You look at healthcare in the US and it's broken down,' he says. 'They spend 18 to 19 per cent of their GDP on healthcare and the life expectancy is decreasing. It's absurd.' China too has a long way to go before its products are trusted in high-end medicine where quality is paramount. 'I always give this example that if you had to implant something in your brain or your eye and if there's a product that's Chinese-made or Swiss-made, I'd rather have the Swiss-made,' Aebischer adds.

In doing so Switzerland would be playing to its traditional strengths. Two out of the world's five largest pharmaceutical companies – Roche and Novartis – are Swiss. Basel, where they are based, is the single-most important city in the global pharma industry, home to thirteen of the world's hundred largest life science companies. Switzerland is the second largest exporter of pharmaceuticals in the world and the industry accounts for almost half of all Swiss exports. These traditional pharma giants have also been snapping up new companies globally, consolidating Switzerland's position in the emerging biotech sector. Bay Area based Genentech, an early pioneer which is widely regarded as the world's first and perhaps most important biotech company, is wholly owned by Roche, the Swiss multinational.

And with these acquisitions Swiss pharma giants are hoping to extend their relevance not just to medicine but to modern life as we know it, where their presence for much of the past century has been ubiquitous but also often inconspicuous. Few would know that Valium, the first blockbuster drug, which from the 1960s through to the 1980s was the most prescribed medication in America, was developed in a lab at Roche. It was the first pill to jump the medical fence and become something of a cultural icon, a symbol of life in the mid-twentieth century and its attendant anxieties which it was meant to overcome. 'Mother needs something today to calm her down,' goes the 1966 Rolling Stones hit 'Mother's Little Helper', 'and though she's not really ill, there's a little yellow pill'. In Woody Allen's 2011 film *Midnight in Paris* when the screenwriter Gil time travels to 1920s Paris, she has an encounter with the writer Scott Fitzgerald's heartbroken wife Zelda who is about to jump into the river. Gil offers her Valium. Zelda: 'I've never heard of Valium, what is this?' Gil: 'Er, it's the pill of the future.' And it was at Sandoz laboratories in Basel, which later became Novartis, that the Swiss chemist Albert Hofmann synthesized LSD during World War II, another compound which would become a cultural force that defined an entire era.

Valium and LSD were not so much drugs designed to cure an illness, which it was hitherto thought to be the entire purpose of pharmaceuticals and the industry built around them, but more like pharmacological lifestyle choices. Expect more of that ahead. Aebischer thinks that there is an entire raft of substances coming our way, not quite medicines, but not quite supplements either, but something in between, serving preventative rather than curative purposes, that would be a part of a normal healthy lifestyle. He sees these products taking the form of semi-complex health promoting preparations that are easily accessible at home, something like Nespresso, the home brewed coffee machines, also a Swiss invention which took the world by storm. Longevity is of course a major focus for well-funded American companies like Altos and Calico, but for Aebischer they're taking the wrong approach, 'they all think pharma,' he says. The future of anti-ageing is continuous low-grade enhancements to everyday lifestyles rather than invasive medical interventions, and this, for Aebischer, is what separates the Swiss from the American approach. 'Let them work on longevity and we'll work on health span,' he says.

7

In 1992, after spending a decade in academia in the US, Aebischer passed over an opportunity to become an endowed chair at Harvard to return to Switzerland. It was not an unusual choice. Many of the best graduates of the Swiss education system find their way into marquee names in academia and industry in the US and UK only to eventually head home as they approach middle age. In that they are different from professionals who emigrate out of other places in Europe and Asia who often leave for brighter prospects across the Atlantic and then never look back. The reason is usually the quality of life in Switzerland which compares favourably to just about any other place in the world. Schools are free, healthcare is excellent, air breathable, and streets safe. Zurich and Geneva, the country's two largest cities, rank among the five most liveable in the world.

Aebischer attributes much of this to the country's relatively sane no dramas politics which has underpinned hundreds of years of political stability. 'Switzerland is the only country that was able to master the ego of politicians because the system is so decentralized,' he says. Politics rarely centres on individuals and supreme executive authority rests with a committee of seven people, known as the Federal Council, instead of a single head of state. The presidency is largely ceremonial and it's not unusual for regular people to not even know the name of the person holding the country's top political office. Swiss politicians, even at the highest level, are known for their relative anonymity and down-to-earth personalities.

In 2014, a picture went viral of then President Didier Burkhalter casually tapping away on his phone while waiting for his train on a platform without a security detail or other hangers-on or even any hint that anyone around him recognized who he was. In 2018, another photo made the rounds of President Alain Berset sitting on the pavement outside the UN building in New York taking notes in between sessions of the General Assembly with passersby not having the faintest idea that the man on the kerb ran one of the twenty largest economies in the world. In another incident, Doris Leuthard, one of seven people on the Federal Council and a former president, was pictured sitting on the stairs of an overcrowded train because she couldn't find a seat on board.

Soon after taking the helm at EPFL, Aebischer went shopping for faculty who fit the same profile as him: Europeans who felt they had got what they wanted out of their sojourn abroad and were waiting for the right opportunity to make their way back home. One of his top recruits was Edouard Bugnion, an ETH graduate who spent eighteen years in Silicon Valley, first as a student at Stanford, then as co-founder of VMware, which would later be acquired by Broadcom for $69 billion, still the largest acquisition in the Valley ever, and then as a top executive at Cisco. He tells me that though he had lived the American dream he didn't think twice before trading it for a spot on the faculty at EPFL, in fact he felt lucky to be able to do so. 'Switzerland is an attractive place to live,' he says. 'It's a country that values its returnees, I think more than others. Other countries are more insular in nature and if you leave then you're outside the system and it's difficult to come back. Switzerland, because we're such a small country, goes out of its way to make it easy for you to come back. So if you have Swiss roots and you have experience elsewhere and you come back, it's something that is valued and appreciated.'

Marcel Salathé, also an ETH and Stanford alum, and who has also taught at Stanford, is another Swiss émigré who couldn't resist the pull of home as the years went by. 'I always felt I have two hearts in my chest, one European/Swiss and the other American,' he tells me. 'The American one is about can-do, pioneering new frontiers, which I absolutely love. And the thing that attracts me to Europe is the more societal aspect, people sticking together when times get tough, proper infrastructure, real social networks, those kinds of things. And eventually I realized that one of them is easier to transport around. I can't take the Swiss infrastructure with me when I go to the US. But I can take that American spirit with me to Europe. And so that's why I said I'm going to go back.' Salathé now heads the Digital Epidemiology Lab at EPFL and is the organizer of Applied Machine Learning Days, or AMLD, a prominent gathering of the AI community in Europe.

This systematic recycling of talent has been to the benefit of the country's technical capabilities, often placing it at the forefront of the state of the art in the world. In 2020, when Covid hit, Bugnion and Salathé teamed up to develop Switzerland's response to the pandemic. Under the leadership of another EPFL faculty member, Carmela Troncoso, and in partnership with collaborators at ETH, their team developed a protocol called the Decentralized Privacy-Preserving Proximity Tracing, or DP-3T, which laid the technical foundation for contact tracing, the system for identifying and notifying individuals who might have been exposed to an infected person. DP-3T would later be picked up by Apple and Google for their joint exposure notification system, known as GAEN, which was incorporated into iOS and Android and rolled out to billions of users worldwide. Switzerland was the first country in the world to come out with an app for contact tracing built on the GAEN framework. The team at EPFL essentially helped establish the global gold standard for contact tracing, with DP-3T influencing not just the system adopted by the big tech companies but also by countries including Austria, Belgium, Croatia, Germany, Ireland, the Netherlands and Portugal who built their contact tracing solutions based on this protocol.

Like trains and universities and healthcare, here too the idea that technologies and systems inevitably reflect the values of the context in which they arise was on display. There were many competing protocols for contact tracing that came up around the world. Singapore's government came out with BlueTrace which was adopted by Australia, the UAE and other countries in the Asian hemisphere. A European consortium launched what was called the Pan-European Privacy-Preserving Proximity Tracing, or PEPP-PT. China had its own national system. All these approaches were centralized, they gathered data from users' phones and stored it in a central database. The DP-3T stood out particularly because of its decentralized and privacy preserving character, both core Swiss values. The system used Bluetooth instead of GPS, which gathered information only about whether people were in close proximity and not about where exactly they were. A phone belonging to someone who had tested positive would send anonymous alerts directly to other phones that had been nearby without uploading this information to a central database, ruling out the possibility that the authorities could maintain detailed logs of infected people and their identities and locations or engage in a data grab that could be used for purposes extraneous to the crisis at hand. This risk proved to be more than just theoretical when it was subsequently revealed that in Singapore law enforcement agencies had used contact tracing data in criminal investigations.

DP-3T's privacy preserving features were partly the reason why big tech companies chose this protocol over competing approaches, with Google's Vice President for Android Dave Burke specifically noting that the Swiss protocol 'gives the best privacy preserving aspects of the contacts tracing service'.

Salathé adds: 'Many European governments did not want to have anything to do with this decentralized approach. They wanted to have a centralized approach. And that really shocked me quite deeply. Eventually, of course, they realized it's not them making the decision, but Google and Apple, which was a shock to them but welcome to the new world I guess. Had the European governments had their way we would not have this privacy preserving contact tracing protocol that we have now. That to me was quite a shock, because I was always thinking, well, Europe, that's where everyone is privacy conscious, and in the US maybe not so much. And the way the dynamic played out was exactly the opposite. The European governments, many of them wanted to have a non-privacy preserving solution where they said privacy means trusting us, whereas the two American tech giants said, no, no, no, no, we're going to build privacy into this. I could not understand what was up and what was down any more.'

The implications of this partnership between companies from

one country, universities from another, and governments across continents go beyond DP-3T. It outlines a model that we're likely to see frequently in the years ahead. Salathé, who was the most quoted scientist in the Swiss media during the pandemic, notes that this experience shines a light on what public policy everywhere is going to look like in the future. 'If I now think ahead of new future crises, I am not expecting public institutions to be able to really solve these for us,' he says.

8

Returnees are not the only ones bringing top-tier expertise from the outside world to Switzerland. The country has an unusually high proportion of foreign academics working in its universities. Four out of every five faculty members at EPFL is an immigrant. 'I don't know how many countries would accept that,' says Aebischer, noting that Switzerland, because of its size, has no option but to tap a global talent pool to stay competitive. 'If you don't recruit from outside, there's no chance.' Much of the country's reputation in the sciences is built on foreign talent and over half of its thirty Nobel Prizes were won by scientists working at Swiss institutions but who were born overseas.

One of the more recent superstar arrivals is Maryna Viazovska, a 40-year-old mathematician of Ukrainian origin who joined EPFL in 2017 and where she is now a full professor. In the spring of 2022, just as Russia began its invasion of Ukraine, where she still has family, Viazovska won the Fields Medal, the discipline's highest honour, making her the second and currently the only living woman to win the award in its eighty-eight-year history. Viazovska, who's work focuses on the most efficient way to pack spheres in a given space, and who was educated in Ukraine and Germany, declined an offer from Harvard to take the appointment at EPFL where her husband, a physicist, is also on the faculty.

She is the latest of a clutch of top-tier mathematicians who have taken up residence in Switzerland which can now claim to have as many as six Fields medallists, all six of them immigrants, who are still active in the discipline. Until recently they had seven, but in 2023, Wendelin Werner, a German-born French mathematician who had been a long-time faculty member at ETH, left for Cambridge. This is an unusually high concentration of elite math talent in such a small geographical area, perhaps the densest in the world, all within a three-hour train ride, given that there are less than fifty Fields medallists who are still alive and even fewer who are still making contributions to the field.

It's not just scientists who have been moving there. Switzerland is not generally seen as a nation of immigrants, a self-image that it actively resists, but it hosts a higher proportion of foreign-born people than countries that are more explicit about immigrant identity, like the US, and comparable to places like Canada and Australia which are actively trying to re-engineer their demographics. One in four Swiss residents was born overseas. Almost a third of the Swiss population are immigrants or descendants of immigrants. The country adds 100,000 immigrants, or 1 per cent of its population, every year.

Their contributions go far beyond science. Much of Switzerland's outlier success in business is also the work of overseas talent. Nestlé, its largest company, which is also its largest employer, and which has been the dominant force on its corporate landscape for well over a century, was founded by Henri Nestlé, a German political refugee. Swatch, the largest watchmaker in a country known for watchmaking which is, in fact, the largest watchmaker in the world, was founded by Nicolas Hayek, an immigrant from Lebanon. Valium, the world's first blockbuster drug, which saved Roche from certain bankruptcy, was developed by Leo Sternbach, a Polish refugee. Beyond business, too, immigrants have left their mark on Swiss cultural and intellectual life. Roger Federer's mother is from

South Africa. Lenin and Trotsky, though by no means Swiss, spent consequential periods of their lives in Switzerland and made this small country relevant to big history that transpired far beyond its borders.

They weren't the only short-term immigrants who left their mark. The most famous example is of course Einstein who came up with the theory of relativity while working at the Swiss Patent Office in Bern before leaving for Princeton. Another short-term resident is Vitalik Buterin, who invented Ethereum while living in Zug, which is where the Ethereum foundation is still based. Zug's status as the unlikely birthplace for the world's second-most important cryptocurrency has made this unassuming city in central Switzerland one of the more important hubs for blockchain technology, sometimes called 'Crypto Valley', a branding that the cantonal authorities have actively leaned into, even allowing its residents to pay their taxes with cryptocurrencies.

The flow of migrants has kept up in part because Switzerland has historically been not just a neutral but also an open and tolerant country. It never really saw the same sort of repression and scapegoating of minorities that until a generation ago was a prominent feature of domestic politics in larger European states. In fact, the country has historically been a destination for minority ethnic groups fleeing persecution elsewhere on the continent, bringing their talents and drive with them.

The earliest, largest, and, from an economic perspective, most important wave happened in 1572 when over 20,000 Huguenots, French Protestants who followed the Calvinist tradition, were systematically slaughtered by Catholics in Paris in what was known as the Massacre of St Bartholomew's Day. The French Huguenots fled to neighbouring countries in droves and many of them settled in Switzerland. The very word 'refugee' originates from this Protestant migration to Switzerland, derived from the French word 'réfugié', which referred specifically to the Huguenots fleeing religious persecution in France. As an ethnic group the Huguenots are known for their work ethic and commercial acumen; John Rockefeller, Henry David Thoreau and Warren Buffett are some of the more noteworthy Huguenot descendants.

In Switzerland, the new arrivals laid the foundation for the industries for which the country is known today: watchmaking, finance and pharmaceuticals. Prominent Geneva banks like Pictet, Lombard Odier, Mirabaud and Bordier are all products of this Huguenot migration. Switzerland has since had a long tradition of sheltering refugees fleeing political violence abroad, accepting large numbers of people fleeing armed conflict throughout the past century, from the wars in Hungary and Czechoslovakia to the more recent conflicts in Ukraine and Afghanistan.

James Breiding, the author of *Swiss Made*, an account of the country's corporate history, notes that Switzerland's much maligned policy of banking secrecy emerged initially from this well-intentioned and principled desire to protect 'the private sphere from state repression'. Political asylees could deposit whatever valuables they had brought with them discreetly without placing their assets at risk of seizure. Private banks like Pictet had initials on their doors rather than their full names and clients could enter and exit using disguised entrances at the back. Hans Baer, heir to the Julius Baer banking dynasty, notes in his autobiography: 'It could happen that a client introduced himself with a bottle of cognac: "My name is Hennessy. I don't want to say more. Here is $300,000." We accepted the money gladly, thankful for the trust placed in us.' Nicolas Hayek, the founder of Swatch, has said that 'the great value of Switzerland is that it provides refugees with physical and financial asylum'.

Switzerland's humanitarian tradition is not just the unlikely origin of its trademark banking secrecy but also the starting point for its status as something of a global capital for international organizations. The first was the International Committee of the

Red Cross, or ICRC, established in 1863, which paved the way for the Geneva Conventions, widely regarded as the beginnings of modern international humanitarian law, and the foundation for 'a modern, multilateral order that puts people first and upholds the supremacy of law over power'. The organization's flag, red cross on white background, is a mirror image of the Swiss one. The ICRC was the first to establish Geneva's international character and the city's cosmopolitanism has snowballed ever since. It now hosts the UN's European headquarters, the organization's largest presence outside New York, and its constituent organizations like the WTO and WHO. Switzerland is also the location of choice for other international non-governmental organizations: the International Olympic Committee is based in Lausanne, FIFA is based in Zurich. This wave of international organizations has been followed by private multinationals and global tech companies. Google's largest engineering hub outside the US is in Zurich, and it is here that Google Maps was developed.

But just because Switzerland has a lot of immigrants doesn't mean it also happens to like them. Mainstream outlook towards the influx of foreigners is, at best, ambivalent and the tension between the economic necessity and sometimes humanitarian obligation of taking in more people and the political preference for preserving the country's demographic character is a recurring and polarizing theme in Swiss public affairs. The sentiment was best captured by the writer Max Frisch who wrote in 1965: 'We wanted workers, we got people instead.' Switzerland has in recent years tightened its immigration rules for asylum seekers and foreign workers. EPFL and ETH, which have more international students than Swiss ones, have tripled tuition fees for overseas applicants and are mulling introducing caps on foreign enrolments. Stemming this flow of migrants though is going to be an enduring challenge. Switzerland, unlike other migrant destinations, receives people mostly from other European nations, three quarters of whom come from within the EU. Its access to the

European single market depends on a reciprocal willingness to adhere to all four EU freedom principles, which includes the movement of people. Given the limited room to manoeuvre, the policy posture so far has been to muddle through, which is unlikely to change for the foreseeable future.

9

Switzerland's international character is not just shaped by returnees and immigrants but also by an unusually large diaspora. One in ten Swiss citizens live abroad, one of the highest ratios of expatriate to native population of any nationality in the world, a cohort large enough that it is sometimes called the 27th canton, or the Fifth Switzerland, a reference to the four distinct linguistic regions – German, French, Italian and Romansch – that make up the country. All this coming and going of natives and foreigners has brought ideas from all over the world to converge on this small stretch of land in the heart of Europe. Some of these émigrés play an outsized role in their adopted countries but are often so well-integrated into their environments that it is not uncommon for their Swiss origins to sometimes fade out of view.

Louis Chevrolet, Hansjörg Wyss and Guillaume Pousaz are good illustrations of three generations of Swiss entrepreneurs who have made their mark abroad. The Chevy is seen as a quintessentially American brand but it has its origins in the decidedly un-American sounding town of La Chaux-de-Fonds in French-speaking Switzerland. It is here that the company's founder, Louis Chevrolet, was born in 1878. The self-taught engineer and world-record-holding racing driver would later move to Philadelphia where he would catch the eye of William Durant, the founder of General Motors, who after being ousted from his own company would team up with Louis to set up a rival car company, Chevrolet, only the third car company to be founded in the US. Louis worked on the very first front-wheel-drive cars and these early innovations helped the company outcompete Ford and GM to in short order become the largest car company in the US. But Louis's life would follow a tragic path. He had a falling out with Durant and as a result left the company and sold all his shares. When his other entrepreneurial ventures failed, he was forced to go back to Chevrolet on unfavourable terms, not as a manager, but as a regular mechanic working on the assembly line, doing menial tasks as an anonymous worker for a company that still bore his name just to make ends meet. Louis died in poverty in 1941 but left an enduring legacy in one of America's most iconic brands. In *La Place de la Concorde Suisse* John McPhee writes that Chevrolet's bowtie emblem is directly inspired by the cross on the Swiss flag: *'n'est pas sans rappeler, de façon stylisee, le pays d'origine du constructeur'*.

There are other cheerier stories. Just as Louis Chevrolet was exiting the scene, another Swiss entrepreneur who would make his presence felt in the US, Hansjörg Wyss, was born in modest circumstances in Bern in 1935. His father sold calculators and his mother was a homemaker. Wyss studied engineering at ETH and then made his way to Harvard Business School where he got his MBA in 1965. He would later make it big as the founder of Synthe, a medical device manufacturer, which he sold to Johnson & Johnson for $20 billion in 2012. And so began his second career as top-tier philanthropist and major behind the scenes force in American politics. Wyss has given over $700 million to Harvard, making this quiet Swiss-born billionaire, who now lives in Wyoming and rarely gives interviews to the media, the largest donor in the school's history bar none. Mark Zuckerberg and Priscilla Chan ($519 million), Ken Griffin ($500 million) and Leonard Blavatnik ($200 million), boldface names that most frequently come up as the school's major benefactors, all rank after Wyss, a name that would be unknown to most of the school's alums. The 89-year-old billionaire has in recent years increasingly found himself in the crosshairs of right wing groups and all shades of conspiracy theorists who ascribe sinister motives to what they consider to be his outsized role in US politics, the new Soros, a generous benefactor of progressive causes who the *New York Times* has called 'one of the most important donors to left-leaning advocacy groups and an increasingly influential force among Democrats'. According to a biography of Wyss written by his sister, his goal is to 'interpret the American Constitution in light of progressive politics'. He has given almost a quarter of a billion dollars to the liberal Sixteen Thirty Fund alone which organizes campaigns on issues like abortion, minimum wage and voter registration. Wyss has also played a transformative role on the Swiss technology scene, directing over half a billion dollars to set up major scientific institutes like the Wyss Center for Bio and Neuroengineering in Geneva, the Wyss Translational Center in Zurich and the Wyss Academy for Nature in Bern. I've lost count of the number of Swiss deep tech entrepreneurs who told me they received funding from one Wyss-related foundation or the other, none of which required them to hand over any equity, making him the single largest individual private donor to Swiss innovation and research.

Leading the current generation of Swiss-born tech entrepreneurs is Guillaume Pousaz, the founder of Checkout.com, a payments processing company, analogous to Stripe in the US, which serves major brands like Netflix and Siemens. When the company raised a billion dollars for its Series D round in 2022, at a valuation of over $40 billion, it became for a while the most valuable startup in Europe. Pousaz dropped out of EPFL and later moved to Singapore to launch his company which eventually re-domiciled to London. The Geneva native grew up in difficult financial circumstances and joined the ranks of billionaires by his late thirties. He now lives in London but still has strong ties to his home country which is where his family office, Zinal, named after a mountain village in Switzerland that marks the end point of a gruelling 31-kilometre mountain race, is based. It is a prolific investor in Swiss startups.

10

The stories of these émigrés were enabled by a thriving culture of entrepreneurship in Switzerland which has made this small country a radical outlier when it comes to business performance. Eleven of the world's 500 largest companies and thirteen of Europe's 100 biggest corporations are Swiss. Switzerland ranks eighth in the world for the greatest number of Fortune 500 companies, which is a remarkable figure, given its minuscule size compared to the others on that list: US, China, Japan, Germany, France, South Korea and the UK. It has the greatest number of Fortune 500 companies globally bar none when measured relative to its population: more than three times the US and fifteen times more than China. And in contrast to other small rich countries like Singapore and Qatar which built their wealth on a relatively narrow economic base, Switzerland has globally known companies across a broad range of industries like banking (UBS and Credit Suisse), food (Nestlé), pharma (Roche and Novartis), commodities (Glencore), insurance (Zurich and Swiss Re), watchmaking (Richemont, Patek Philippe, Hublot and Rolex), robotics (ABB), hospitality (Ritz), and, of course, chocolates (Lindt & Sprüngli and Barry Callebaut). This is remarkable outperformance by a landlocked country with little to no natural resources.

'Switzerland has always been strong in using its Swissness as a really reputable brand internationally,' says Oliver Heimes, a partner with Lakestar, a venture fund based in Zurich. 'There is a premiumness associated with Switzerland which has historically made its companies less susceptible to low-cost competition from abroad. It probably wouldn't be that hard for a Chinese company to come out with a product that, technically speaking, is the same as what's on offer at Rolex or IWG and massively undercut them on price. But that simply wouldn't work. China exports 30x more watches than Switzerland. And yet in 2023, Switzerland's watch industry made $28 billion in revenue while the Chinese only made $2 billion. The difference is that while the average price of a Chinese watch is $4, for a Swiss watch it is $1,679. That brand advantage is hard to beat. 'Swissness has become a brand in its own right,' writes James Breiding, the author of *Swiss Made*. 'And this may be the country's most precious and enduring comparative advantage.'

A new generation of consumer brands have been coming out of Switzerland which tap into these positive associations, the most prominent of which is ON, an athletics sportswear brand which since its launch in 2010 has become one the fastest growing lifestyle brands globally. The company, backed by Roger Federer, who is both an investor and a brand ambassador, is now worth over $11 billion. ON's fortunes have been buoyed by a blurring of the line between athletics and everyday life in the wider culture. 'If you think about what has happened over the last few years, it's that sports is the new uniform,' the company's founder, David Allemann, told me. 'So while the prime archetypes of fashion in the past have been more kind of the military uniform and the classical jacket and the coat, which also came from kind of military archetypes, now it's the hoodie and the track pants and the tights and the sneakers.' Allemann says that Switzerland has everything to do with the company's success: it's a small country, it's easier to stand out and reach a tipping point compared to larger markets like the US where a startup like his would have easily been drowned out by the noise. And then there's brand Switzerland, the inherent advantage of every Swiss company. 'We are building a premium sports brand and the emphasis is on sports and on premium,' Allemann tells me. 'And I think this premiumness is very much associated with Switzerland. Engineering is very much associated with

Switzerland. Whether it's watches or railways. So kind of premiumness and engineering come together very much in the DNA of the Swiss brand. And that's why we put a flag on our shoes and proudly write Swiss engineering next to it because that's our DNA.'

If premiumness is one aspect of brand Switzerland, trust and discretion are the others. Patrick Aebischer points out that quality and reliability have become core elements of Swissness: 'Switzerland is a kind of trust, reputation is the brand,' he says. This has made it a hub for companies that play a big role in what Eduard Bugnion, the professor at EPFL, calls 'the trust protocols of world commerce'. Their presence is ubiquitous but also hidden given that their entire business model is based on secrecy. The poster child for this sort of company would be SICPA, a hundred-year-old family business which is not listed on the stock market and does not allow any outside shareholders. It first came up as a provider of inks for identity documents, passports and protected materials, including securing most of the world's banknotes including major currencies like the dollar, euro and the Swiss franc. It eventually branched out into adjacent areas like tracing counterfeit goods and digital seal technology for registries. The company is a major sponsor of the 'unlimitrust campus' in Prilly, an industrial area northwest of Lausanne, a vast complex that serves as an incubator for startups that 'promote the economy of trust', an ecosystem that is supported by the cantonal government and EPFL.

A more recent entrant in this space is ID Quantique, a Geneva-based quantum technology company which helps organizations keep their data safe using quantum-safe network encryption. Since 2007, ID Quantique's technology has been used to protect the voting process in Geneva, securing the connection between the central ballot counting station and the government data centre. The company's founder, Grégoire Ribordy, tells me that quantum technology is a lot more advanced and market ready than people generally think and that its relevance is growing in the face of the need to protect against 'reactive vulnerability': the idea that much of the world's sensitive data is currently protected by conventional encryption methods, which would be easy to crack with quantum computers which are just around the corner, so organizations need to prepare ahead of time and upgrade the encryption levels of their existing stores of data in anticipation of the arrival of these more powerful systems. 'I would say we were fifteen years too early,' Ribordy tells me, but now things are taking off in the field of quantum security with high-profile data breaches like the Sony hack revealing that by the time organizations realize they're vulnerable to newer threats it's already too late, so future-proofing systems is moving higher up on the corporate agenda.

Switzerland has a history of growing tall companies and has a healthy ecosystem of new startups. But none of them have scaled to the level of some of their globally known predecessors like Nestlé and Novartis. Why? I put this question to Anja König, the Global Head of the Novartis venture fund, based in Basel. The limitation, she says, is mostly the financing. Swiss companies have good access to the smaller amounts of money needed to get going but don't have robust access to the larger amounts of money in the public markets that are needed to grow. Two reasons. First, the absence of a European equivalent to the Nasdaq, a technology focused stock market that has less stringent requirements for listing, so newer ventures can raise public funds more easily. Europe's stock market landscape is extremely fragmented with forty-one different exchanges compared to just three main exchanges in the US, which dilutes the already limited pools of risk capital across too many different competing centres. 'The single most important thing that Europe should think about is the European Nasdaq idea to be able to keep companies in Europe and build them up sustainably and somebody needs to take the leadership,' says König. 'Switzerland might as well try because they have a pretty strong banking sector.' The second reason is that Switzerland is not a member of the EU, which limits its access to European funding mechanisms, the biggest of which is the European Investment Fund, or EIF, which invests in venture funds on the continent. But that money comes with strings attached: most of these EIF funds must be channelled to EU companies. Since Switzerland exists outside of that context its companies can't tap these funds and instead compete for the relatively limited funds that venture capital funds have earmarked for outside investments, which happens to be a very competitive field. 'This means that Swiss companies are directly competing with American companies,' König tells me. 'Now Swiss companies are very competitive for many reasons within Europe, but competing with the US is not that easy. So I think this is the single most important issue that Switzerland needs to fix: to clarify its relationship to these European funding mechanisms.'

CHAPTER SEVEN

The New Mittelstand

'Silicon Valley is good for high tech, but Germany is good for deep tech.'

I

Germany has a long and storied history in aviation. It begins with Otto Lilienthal, a nineteenth-century German aviation pioneer known as the 'flying man', who developed some of the earliest gliders capable of sustained and repeated flights. Lilienthal's company, Maschinenfabrik Otto Lilienthal, was the world's first aeroplane company, and his Lilienthal Normalsegelapparat, German for 'normal soaring apparatus', was the world's first aeroplane in series production.

Lilienthal flew over 2,000 flights until he died in a fatal crash in August 1896. He was only 48. His last words, also inscribed on his tombstone, were *'Opfer müssen gebracht werden!'* ('Sacrifices must be made!'). Lilienthal's untimely demise prevented him from achieving his goal of powered flight. But his bold experiments were a direct inspiration to the Wright brothers who did manage to be the first to achieve that feat. Wilbur Wright said of Lilienthal:

The New Mittelstand

No one equaled him in power to draw new recruits to the cause; no one equaled him in fullness and clearness of understanding of the principles of flight; no one did so much to convince the world of the advantages of curved wing surfaces; and no one did so much to transfer the problem of human flight to the open air where it belonged . . . he was without question the greatest of the precursors, and the world owes to him a great debt.

Germans made other major contributions to early aviation. The V2 rocket, developed by Wernher von Braun during World War II, was the first man-made object in space. Von Braun would later be captured by Allied troops and moved to the US at the end of the war as part of Operation Paperclip, a secret government plan to bring over a thousand German scientists to the US to work on military technologies.

The former German nationalist took little time to ease into his new life as an American patriot. Von Braun is now regarded as the father of the US space programme, the one man most responsible for bringing the US into the space age. He designed all the early rockets that mattered: the Jupiter-C that powered Explorer-1, the first American satellite; the Redstone that powered Project Mercury, which put the first American astronaut into orbit; and the mighty Saturn-V that powered the Apollo missions, which put humans on the moon.

Jet propulsion in general is a German invention. The world's first civilian jet aircraft, the Heinkel He 178, developed in 1939, and the world's first jet fighter aircraft, the Messerschmitt Me 262, developed in 1941, were also made in Germany. But after these early triumphs, Germany largely ceded its pre-eminence in aviation to the US. Today, four out of five of the world's largest aircraft manufacturers and all five of the most valuable space companies are American.

Lilium, an aerospace company based in Munich, which takes its name from Otto Lilienthal, wants to restore Germany's standing in aviation. It was founded in 2015 by four graduates of the Technical University of Munich (TUM), the country's premier aerospace engineering school. Lilium makes compact electric jets, also known as eVTOLs, or electric Vertical Take-off and Landing Aircraft.

Lilium is riding the next big wave in aviation. At present, air travel is typically used for long journeys, between cities and continents. eVTOLs promise to make flying a viable option over shorter distances, within cities and suburbs. These small electric planes can take off and land vertically and hover around like a helicopter. Some are nimble enough to roam around freely in dense urban environments, capable of picking up and dropping off passengers on rooftops or even crowded streets. The idea is to change air travel from an elaborate ritual that involves tickets, check-ins, airports and runways to something that feels more like hopping on a bus.

At any given moment there are anywhere between 8,000 to 20,000 planes in the sky. We can expect those numbers to climb sharply as eVTOLs roll out. This will be a step change in what it means to fly.

The Lilium Jet has a fuselage that is not much larger than a standard SUV. It has room for up to six passengers and can fly 175 kilometres on a single charge, about the same range as an average electric vehicle, with plans for longer range aircraft in the future. It emits zero CO_2 and is quieter than regular planes. With its smooth curves, minimalist interior and touchscreen controls, Lilium very much lives up to its billing as the Tesla of the skies.

'We wanted to prove that it can be done here,' Daniel Wiegand, the company's young and charismatic co-founder, told me. 'If Elon can build SpaceX in the US, why can't we build this here?'

That idealism has been severely tested. When Lilium went public in 2021 it was valued at $3.3 billion, one of the biggest stock market debuts by a German tech company ever. It has since lost 80 per cent of that value. The company's most public crisis came in 2023 when its stock price dropped below a dollar, raising the imminent prospect of a delisting by Nasdaq. It managed to avert disaster by securing a last-minute $175 million cash infusion from Tencent. Scepticism around the company has also grown because of delays with the large-scale roll-out of its aircraft.

In the face of mounting challenges, Wiegand, who founded the company on the premise that big things *can* happen in Germany, has sometimes found himself wondering: *can* big things happen in Germany? 'In retrospect, I think in some sense we prove that it can be done, because we still exist, from another perspective in some areas we proved that it's a challenging environment,' he says. 'There's not enough money, there's not enough growth capital, there isn't enough of a risk-taking culture. The employees in Germany are more conservative and not everybody dares to swap a comfortable job in a good company with the risky adventure of a startup.'

Lilium sums up both the progress Germany has made in creating a more supportive environment for high-risk ventures but also the challenges that still confront its most ambitious companies. The most tired question in all European industry is: why hasn't the continent produced anything like an Apple or a Google; a large, globally relevant tech company that is known everywhere from Chile to Japan? And the answer in so many different ways comes back to what Wiegand is talking about above: capital and culture. The environment is too placid, there isn't enough money for risky ventures, there aren't enough entrepreneurs chasing bold new ideas. Lilium was among the first wave of companies shaking up that status quo.

'Lilium has a shot at being extremely disruptive,' says Francesco Sciortino, founder of Proxima, a fusion energy startup in Munich. 'It turns out that the idea is more complicated and more expensive to bring to fruition than some people had expected at first, but still it's an amazing company.'

Lilium may yet change what it means to fly. But it can credibly claim to have already changed startup culture in Munich. When it first came up a decade ago there wasn't much else going on. Fast forward ten years and the city can now legitimately claim to be the deep-tech capital of Europe, spinning out new ventures on everything from robots to rockets to autonomous cars. 'Many of those founders said to me, we've done it here because we've seen that you're still alive and kicking,' says Wiegand. Others seemed to agree. 'This is very good from an ecosystem perspective,' says Sciortino. 'It makes us all more ambitious.'

2

Germany emerged as a unified political entity in its modern form when Otto von Bismarck proclaimed the new German Empire from the podium of the Hall of Mirrors at the Palace of Versailles near Paris at the end of the Franco–Prussian war on 18 January 1871. The newly anointed German Emperor, Kaiser Wilhelm I, imposed heavy reparations on the defeated French. Almost 5 billion francs, or a quarter of the entire French GDP at the time, was paid in indemnities from France to Germany over a period of less than three years. Some of this capital was used to fund new ventures.

'The Kaiser had a lot of money from winning the French War,' says Rafael Laguna, the founding director of the Federal Agency for Disruptive Innovation, or SPRIND, in Leipzig. 'They used it to help professors found their companies; the chemicals industry was created, the pharma industry was created, the car industry was created, this was all in those twenty, thirty years. Millions of Deutsche Mark which would translate to billions today was poured in by that government.'

The Kaiser's massive investments kicked off what is known in Germany as *Gründerzeit*, or Founder's Era, a period of transformative economic boom which saw the creation of dozens of new industries and hundreds of enterprises. Virtually all the companies that are synonymous with German industry even today – Siemens, Bayer, BASF, Deutsche Bank, Allianz, Bosch, Braun, BMW, Mercedes – came up as a direct consequence of the *Gründerzeit* in the late 1800s.

These investments would set Germany up to compete in virtually all the industries built around the new technologies of the early twentieth century. Karl Benz and Gottlieb Daimler invented the world's first automobile in Mannheim in 1886. Karl Ferdinand Braun invented some of the very first versions of the television in Strasbourg in the 1890s. And Konrad Ruse built the world's first fully functional programmable digital computer, the Z3, in Berlin in 1941.

It was largely thanks to the *Gründerzeit* that a hundred years ago Germany sat comfortably alongside the US and UK as one of three most technologically capable nations in the world. But while the country excelled in developing and commercializing technologies of the early twentieth century – radio, television, cars, planes – it barely gets a mention when it comes to the technologies of the late twentieth century: computers, internet, mobiles. Why?

Depends on who you ask. In Germany, I got two very different and very contradictory answers. The first is the obvious one. Germany was on the receiving end of a whole lot of history that went down in the previous century. It was the principal theatre for all three major conflicts that defined that era: two hot wars and one cold one. The country was simply too overwhelmed by geopolitical events to focus on technology and industry. The tumult had a knock-on effect on the German psyche: 'You can do worst-case scenario planning, but no one in the world has ever actually seen this worst, worst, worst case becoming real. But Germany has, right?' says Christian Vollmann, founder of C1, a green energy startup based in Berlin. 'Germany has seen the absolute nightmare, the absolute worst case, become reality and this is what then after that has shaped our perception of risk, we are the only ones who know that the worst-case scenario can actually happen.' He adds: 'I do think that it's holding us back. Germans tend to be sometimes irrationally risk averse, it's not rational any more, that's a big problem if you talk about new technologies because they are inherently risky.'

The other, diametrically opposite explanation is that Germany is not so much psychologically scarred as it is 'fat and happy'. The country went through a lot, took it all in its stride, and came out on top. At the end of World War II, Germany had lost 10 per cent of its population, 40 per cent of all housing, a third of its industrial capacity. Its currency, along with the wider economy, virtually collapsed. Today it is the largest economy in Europe, fourth largest economy in the world, and the third largest exporter globally. The country recovered from war and dismemberment within one lifetime because the postwar generation was willing to roll up its sleeves and get to work. But, some say, the generation after has gone soft and complacent.

'Why are we lagging behind? It's because we're fine,' says Rafael Laguna. 'I mean it's a rich country, right? Gen Z, they're not like the kids eighty years ago where Germany was destroyed. I grew up with this ethical framework of creating value for society, working, getting up at six and doing your thing. They say, look, I've got this one life. I can work for twenty-four hours and have enough money for a week, the rest I can do what I really like doing. And of course, that's not the spirit that you need to create disruptive industries.'

Whatever the precise reasons for Germany's relative decline in building and commercializing new technologies, the net effect is it now sees itself as a massive industrial power that has somehow lost its way in the digital age. A century ago, when the US produced Ford, General Motors and General Electric, Germany responded with BMW, Benz and BASF. But even as the US has produced Apples, Googles, and OpenAIs decade after decade since then,

Germany has struggled to keep pace. And now it wants that to change.

3

The institutional vehicle to effect that change is the Federal Agency for Disruptive Innovation, or SPRIND, established by the German government in Leipzig in 2019. The agency has the explicit mandate to promote and fund disruptive innovations or, if you will, to kickstart a second *Gründerzeit*. Rafael Laguna, the agency's founding director, tells me that the agency is loosely modelled on DARPA, revised and updated for the twenty-first century, and minus the military bits.

The agency's main role is to plug the funding gap between research and commercialization of new technologies, which Laguna calls the valley death. 'I think we're doing pretty good science,' he says. 'Basic research? Great. But we're not doing well translating that into industries.' German science is indeed in good shape. The country ranks third behind the US and UK among nations that produce the most Nobel laureates in the sciences. In per capita terms it outranks even the US.

Laguna has a war chest of a billion euros. He's sort of like a benevolent venture capitalist who doesn't need to take equity in the companies and projects in which he invests. 'We don't need to create ROI,' he says. 'But we need to create returns for the people and we need to create industries that then do their thing in our country, so it's a few levels above that.' DARPA-like agencies are of course now everywhere; Sweden has Vinnova, NATO has DIANA, Japan has Moonshot R&D. But SPRIND is different in that its geographical reach isn't limited to Germany or even Europe. It spans the globe. 'Whoever has the best idea, we will support,' says Laguna.

DNA origami is one of the technologies that has benefited from

SPRIND funding. DNA, the basic building block of life, can be folded, a lot like miniature paper, to create nano-scaled structures in two and three dimensions. It has been used for over two decades to make nanoscale art, sculptures, illustrations and typography a thousand times thinner than a single strand of human hair. Now this approach to nanoengineering is entering practical applications.

DNA origami is being used to make intricate structures like scaffoldings, machines and robots on an atomic scale. SPRIND has funded two Munich-based biotech companies that are bringing this technique into the mainstream: Capsitec, which makes molecular 'cages' or 'traps' that can snare viruses and neutralize them, and Plectonic, which develops nanorobots, called LOGIbodies, that can deliver drugs like anticancer medication specifically to the cells where they are needed. 'The underlying technology is a platform technology that will allow for many, many more applications,' says Laguna.

Germany's biotech industry is having a bit of a moment. Much of that is because BioNTech, the company which developed the Pfizer vaccine, the first jab to come out for the coronavirus, is based there. mRNA, or messenger RNA, the technology on which the Covid vaccine was based, is a novel approach to drug design which instructs cells to produce specific proteins for therapeutic or preventative purposes. Vaccines are just one application. mRNA can be used to develop drugs for a variety of therapies, everything from fighting cancer to gene editing.

BioNTech, founded by a husband-and-wife team of scientists, both of whom are immigrants from Turkey, is now worth over $20 billion, the most valuable company to come out of Germany in the twenty-first century. The company's success has propelled all three of its founders into the ranks of billionaires and transformed the fortunes of Mainz, the city where it is based.

Mainz, a small city of only about 200,000 residents in the western state of Rhineland-Palatinate, last saw the limelight 600 years ago. It was here that Johannes Gutenberg invented the printing press in 1440. The city had since lapsed into relative obscurity and spent much of the past three decades saddled with over a billion euros of debt. BioNTech, which paid €3.2 billion in tax in 2021 alone, flipped the city's finances virtually overnight. Mainz now runs a billion-euro budget surplus. BioNTech's economic reverberations have even been felt far beyond its hometown, on the level of the national economy. In 2021, this one company was responsible for nearly a fifth of the growth of the entire German GDP, an effect that economists say is virtually unprecedented.

And fair enough. BioNTech's success was in no small part enabled by the German government. It has received over half a billion dollars in state support since it was founded in 2008. Its main competitor, US-based Moderna, which made the other widely used mRNA vaccine, has also received public funds to the tune of billions of dollars. In fact, it was a $25-million early grant from DARPA that put Moderna on the path of developing its own mRNA platform. 'The US was lagging behind and only because of DARPA Moderna was created,' says Laguna. 'Otherwise it wouldn't exist and BioNTech and CureVac would have a monopoly.' For Laguna, these examples prove the founding thesis of SPRIND, that sometimes all it takes is for the government to put its thumb on the scale to get a good thing going. 'I think Germany is improving and we see that in numbers,' he says. 'I'm just saying if we put enough chips on the table some of these things will work.'

4

The US is the world's largest economy. Germany, the third largest. But US companies operate on a completely different scale compared to their German counterparts. Fifty-nine of the world's hundred largest corporations are American. Only three are German. This is sometimes taken as evidence that German businesses are less competitive than those in the US. But the disparity is better explained by differences in how large-scale economic activity is organized in these two countries.

The US economy is dominated by large publicly owned corporations like Amazon and Walmart. The concentration of economic power in these companies can be striking. The top 1 per cent of US corporations own 97 per cent of all business assets. The top 0.1 per cent own 88 per cent of all business assets. These large companies account for two-thirds of all US exports.

In Germany things are different. They have big, fat corporations too, like BMW and Siemens. But these are the exception rather than the norm of how large-scale economic activity is organized. The backbone of the German economy consists of over 3 million small and medium-sized firms, known collectively as the Mittelstand. These private, family owned firms typically employ fewer than 500 people. Individually they are modest; in aggregate they are formidable.

Mittelstand companies produce half of Germany's entire GDP, employ over half of its labour force and generate two-thirds of its exports. It is primarily on the backs of these companies that Germany has built its reputation as a global export powerhouse. Germany is the world's third largest exporter, behind the US and China. That ranking seems impressive, but it understates the scale of the country's achievement. Germany's population is a fraction of that of its competitors; the US has 4x more people and China 20x. In per capita terms, Germany exports 4x more than the US and 10x more than China. And it's largely the Mittelstand firms that have made German industry among the most competitive in the world.

The US economy is dominated by a very small number of large companies; the German economy is driven by many small and medium-sized firms. American firms are concentrated in big urban centres: New York, San Francisco, Boston. The Mittelstand is strewn across hundreds of obscure little towns all over the German countryside. All of which goes to explain why, even though the US and Germany are both rich countries with highly successful economies, their companies are completely different in terms of their international profile. It's not just about size. It's also about what a typical company in these two countries does that influences their visibility.

Big American corporations typically make lots of products that are sold to lots of customers in lots of places. Consider 3M. The Minnesota-based company, founded over a hundred years ago, makes 60,000 different products, everything from sticky notes to thermal insulation to dental implants. Its products reach a quarter of the world's population, and an average person comes across a 3M product a hundred times a day.

But where American companies go broad, German companies go deep. Mittelstand firms are known for their maniacal focus. They do one thing and they do it really, really well. They dominate tiny global niches by making continuous gradual improvements to highly specialized products and rack up massive operational efficiencies along the way, a process known as incremental optimization.

These businesses can be almost absurdly specialized. Take for instance Wafios AG, a 130-year-old company based in a small town south of Stuttgart. It makes machines that bend wires into different shapes. That sounds incredibly simple. But it's incredibly hard when it needs to be done on a very small or a very large scale. Those machines produce small but essential components – springs, coils, clips and clamps – used in almost every industry, from cars to planes to phones to medical devices. Demand adds up. Germany has millions of other little firms just like Wafios tucked away in small towns across the country which make that one thing that is crucial to the work of lots of different industries. And these companies thrive by making that one thing better than anyone else in the world.

Hermann Simon, a well-known management professor, has called these companies 'hidden champions': small, highly profitable companies which practically prop up the German economy while staying relatively unknown to the wider public. While Germany has produced relatively few of the world's largest headline-grabbing corporations, it has over a thousand of these hidden companies which rank among the top three in their respective niche markets worldwide, the highest concentration anywhere in the world.

It's not just about bending wires. Germany also has plenty of hidden champions in deep tech. Take for instance Herrenknecht. It makes tunnels. Making holes doesn't sound quite as glamorous as sending people to Mars. But Herrenknecht surpasses Elon Musk's other venture, The Boring Company, in just about every measure. The Boring Company has been around for almost a decade. In this time, it has completed only one major project: a 2.4-mile tunnel in Las Vegas. Herrenknecht has been in business for four decades during which it has completed over 3,600 projects all over the world.

The company has made tunnels for subway networks in New York, Moscow and Doha. Its crowning achievement is the Gotthard Base Tunnel, a 35-mile-long tunnel that passes under the Swiss Alps and connects Zurich with Milan; a marvel of engineering that is the world's deepest traffic tunnel as well as the longest. Herrenknecht, based in Schwanau, an obscure village in Baden-Württemberg, is currently working on other massive global infrastructure projects, including a tunnel under the Panama Canal and one for Britain's second high-speed railway line.

Electro Optical Systems, or EOS, is another hidden champion in deep tech. The company, founded in Gräfelfing, a little-known town south of Munich, makes 3D printers for industrial applications. Its machines, which can cost up to $1.6 million, are installed in 3,000 locations worldwide.

EOS was founded in 1989 by Hans Langer, a German physicist who studied lasers at the Max Planck Institute for Plasma Physics, one of Germany's top research institutes. He tells me that after finishing his PhD he was about to leave for a post-doc at Harvard in 1980 with a career in academia in mind when a professor convinced him to change tracks. 'He said, Hans, don't do this, there are too many of your kind, go into industry.' The advice turned out to be prescient. Langer is today the only person in the world to have made a billion-dollar fortune in 3D printing.

Langer is full of war stories about his five decades in the 3D printing business which seems to suggest that success in this industry is as much about hiring good lawyers as it is about having good engineers and managers. IP wars are rampant. Langer says that he was able to outcompete much bigger publicly traded rivals abroad by tapping into Germany's long tradition of engineering excellence. 'Silicon Valley is good for high tech,' he says, 'but Germany is good for deep tech.'

It was a very different message from what I was hearing elsewhere in Germany where the mood seemed subdued, tinged with a lot of navel-gazing and self-doubt about whether German industry, a giant of the industrial era, was now trapped in an almost Sisyphean odyssey to play catch up in the digital age. Langer was bullish. He thought the country was doing just fine and that the state of collective despair that I had witnessed was just Germans being Germans.

'I'm German by birth, but I'm not German by mindset,' he said. 'I have an American mindset.' Langer talks about how his four-decade old company, ancient by US standards, is still aggressively pursuing new market opportunities in frontier industries. It is the world leader in 3D printed rockets. SpaceX, Blue Origin, Relativity, all the big players are his customers. 'Today we own more or less 50 per cent of the space industry,' he says, 'we are number one in space, this is our fastest growing business.'

Langer has turned down multiple, multi-billion-dollar acquisition offers so he can keep the company in the family. In our discussion about the future of EOS, he talks at length about succession planning and just shakes his head at questions about going public. Autonomy matters. His daughter, Marie, now runs the company and his son, Uli, runs the venture arm.

This is par for the course for other Mittelstand companies. Wafios and Herrenknecht, both family owned businesses, have also reportedly fielded many acquisition offers, increasingly from China. Few sell. These companies are seen by their founders less as assets to be bought and sold and traded in the public markets and more as family heirlooms to be passed from one generation to the next.

This has its advantages. Unlike managers in public companies, Mittelstand owners don't chase quarterly revenue targets, nor do they have to concern themselves with the vagaries of the stock market. And thus, their planning horizon often spans multiple generations. This long-term thinking goes some way to explaining the longevity of these enterprises, some of which predate the founding of Germany as a modern nation-state.

5

Herrenknecht and EOS show that Germany's hidden companies can be every bit as cutting edge and globally competitive as the fiercest 'move fast and break things' variety of tech startups coming out of the US. And because of this the Mittelstand model is often held up as the antithesis and a viable alternative to the Silicon Valley way of doing things. The differences are sharp.

Mittelstand companies are known for their frugality and rejection of outside investment. Costs are pared to the bone. The Albrecht brothers, founders of the discount supermarket chain Aldi, which also owns Trader Joe's in the US, created a fortune exceeding $50 billion. They kept paper-based accounts late into their careers using stubs of old pencils almost too short to hold. They once reprimanded architects designing a new store for using paper that was too thick.

In the Valley, founders are all too comfortable running wild business experiments with other people's money and can sometimes rack up expenses that can only be described as gratuitous. Quibi, a short-form video streaming service founded in 2018, raised a staggering $1.75 billion in funding from top-flight investors like Goldman and JP Morgan. It burnt through most of its cash on expenses like a $5.6-million Super Bowl ad before going out of business within six months.

While US tech founders enjoy rockstar celebrity status and are among the most photographed people on earth, Mittelstand owners are notoriously reclusive. The German public knows next to nothing about Dieter Schwarz, the country's second richest man, who owns the Lidl supermarket chain, and has a net worth of $50 billion. There are only three pictures of him in circulation, one of which is in black and white. He once turned down a medal for entrepreneurial achievement by the state of Baden-Württemberg because he didn't want to be photographed.

But even if Mittelstand owners keep a low profile and engage in few, if any, public relations activities, they enjoy a generally favourable reception from the press and the public. Where the US tech community often finds itself wading from one PR crisis to the next, Mittelstand companies are usually seen as the central pillar of the small communities in which they operate. 'You'd rather be seen hitting your wife and child than talk badly about your company,' *Süddeutsche Zeitung*, a prominent news outlet in Germany, once wrote.

The state too looks kindly upon them. Germany's inheritance code grants big exemptions to owners who keep things in the family instead of selling their businesses. All of which is to say that the Mittelstand model is much admired at home and is seen as worthy of emulation abroad. 'Germany's niche companies are a model for life after globalization,' wrote Adrian Wooldridge in a recent op-ed.

But might the Mittelstand model, the very foundation of the country's wealth in the industrial era, be precisely what is holding it back in the digital age?

'Imagine how much value is locked up in a few people, right?' says Herbert Mangesius, a venture capitalist based in Munich. The fact that such a large proportion of the country's productive capacity is in the hands of a few family businesses has led to an enormous concentration of wealth at the very top of the economic ladder. If in the US it is tech moguls who dominate the ranks of the country's richest, in Germany it is unquestionably the Mittelstand owners. Many are among the richest in the world.

Those who see large public companies as inherently at odds with the common good often have a more sanguine view of small and medium-sized enterprises. But this model can produce economic outcomes that are even more skewed than those wrought by large corporations.

The stock market, for all its faults, serves an important redistributive function. Anyone can buy stock in a public company. If the stock does well, they do well, and wealth finds its way to a lot of people inside and outside the company. Since going public, Microsoft has created eight billionaires and 12,000 millionaires from its employees. The Facebook IPO created a thousand millionaires overnight. The same can't be said for Aldi and Lidl.

When private companies do well, only their owners do well, and their success does not lift the fortunes of those around them. When the economy relies disproportionately on private companies instead of public ones, as is the case in Germany, the fruits of their success are less widely distributed in society.

Germany is often seen by outsiders as a somewhat egalitarian country. The absence of big corporations reinforces this perception. But the fact is that it has one of the highest levels of wealth inequality among Western capitalist countries. And that is a direct consequence of the German economy's structural reliance on the Mittelstand.

A recent working paper by the IMF on wealth concentration observed that German households 'lack access to the German corporate equity stock as most of the corporate net wealth is concentrated in privately-held firms'. A study by Berlin-based economics institute DIW has shown that the richest 1 per cent of Germans own a third of the national wealth, which is just about the same as in the US, the country often held up as something of a gold standard for an unequal society. And thus the SME-based Mittelstand model is not driving equity outcomes dissimilar to their market-oriented corporate counterparts.

German super wealth

Germany has the world's third highest concentration of super-wealthy people, defined as individuals with at least $50 million.

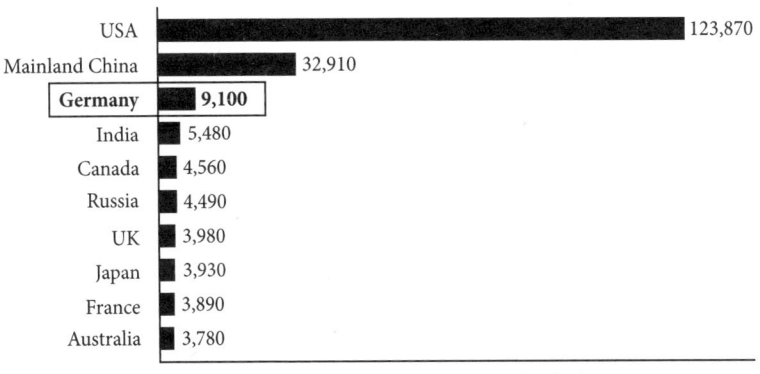

■ Number of ultra-high net worth individuals, 2023

Source: UBS Global Wealth Report (2023)

It is widely understood that Germany is a rich country. But a more accurate statement would be that rich Germans are among the richest in the world while the average German is among the poorest in Europe. Germany ranks behind only the US and China among countries with the most billionaires, and the most ultra-high net worth individuals, classified as people with more than $50 million in assets. And yet it is near the bottom of the rankings of average household wealth in Europe, placing below Slovakia and just above Greece.

Inequality isn't just a bad social outcome. It's also bad for business. An economy dominated by private companies tends to

European household wealth

The average German household ranks below Solvakia and just above Greece.

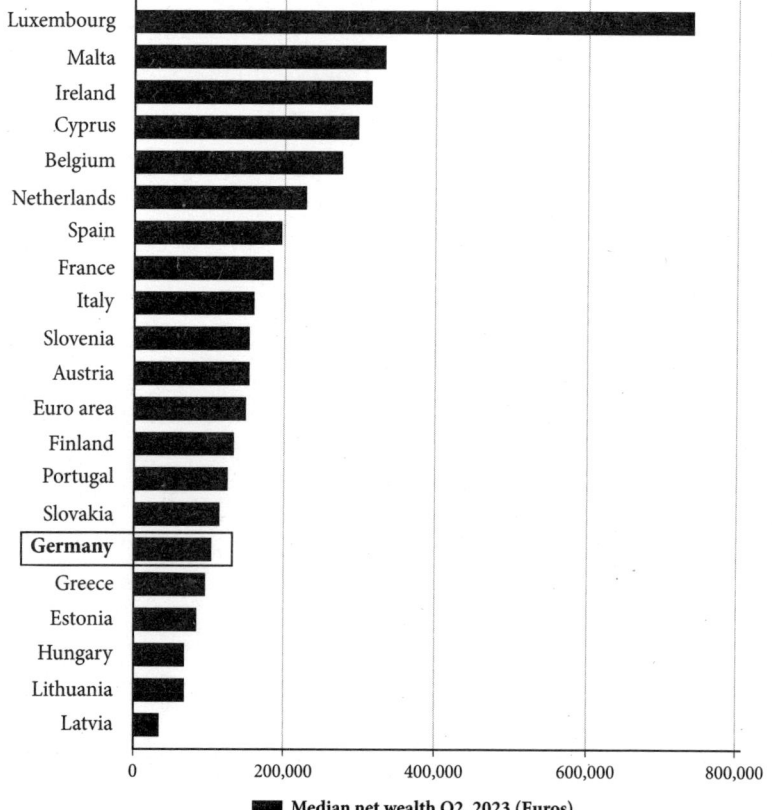

Median net wealth Q2, 2023 (Euros)

Source: ECB Distributional Wealth Accounts (2023)

produce fewer new ventures. In the US, the way new companies are formed is that when one company succeeds the employees and stockholders who held equity in that company then go and launch or invest in other companies – the virtuous cycle of success breeding more success. But when companies are kept private indefinitely and all the equity is with their owners, then wealth is less widely distributed and there's less risk capital circulating in the system. And that's largely what has happened in Germany.

6

The stock market performs the important function of distributing wealth from company owners to others. But it also performs the equally important function of giving company managers access to external sources of capital that can help the company grow. In the absence of this auxiliary capital, companies just don't grow fast enough or big enough to be able to compete with better capitalized overseas rivals.

For a typical Mittelstand company, growth is largely a function of its profits, or its ability to raise debt, which in one way or another still comes back to being a function of company assets. Herrenknecht has profits in the range of $50 million a year. That's a decent sum. But its orders of magnitude lower than the mountains of cash at the disposal of public and venture backed companies in the US and China.

The Boring Company has raised a billion dollars. It is worth $7 billion on the public market and can expand even when taking enormous operating losses. That massive capital advantage may make the fact that Herrenknecht has completed 3,600 projects and The Boring Company just one entirely irrelevant to who prevails in the end. Herrenknecht, once the largest manufacturer of tunnelling machines in the world, is now not even among the top three, all of which are in China.

'You need massive scale in order to compete against that manufacturing, cost and technology development of big corporations, especially the Asian players,' Herbert tells me. 'And most of these Mittelstand companies, they lack the scale.' These are not scenarios for the distant future. Germany has in recent years been losing one industry after another to Chinese rivals. 'The quality on the machine building side of China is getting better and better and better,' Herbert adds. 'And in some industries where we were lagging, like battery equipment, China and Asia are much, much more advanced than we are because we didn't pull through. They are cheaper and better, period. So we've lost that one.'

There are other ways in which this way of doing business can seem outdated. Being hyper focused in one small niche has one big benefit: it's hard for others to compete with you. But this strategy of building slightly better mousetraps also has a major drawback: you can't change quickly when the industry changes. Which is a problem in a world where industries are changing faster than ever before. If a company's entire business model is to make that one perfect piece of coil that goes into every single combustion engine in every single car in the world, it'll do just fine, all the way until petrol cars are replaced by electric ones. And then it's game over.

Incremental optimization makes these companies really good at doing what they're already doing, but really bad at coming up with new ways to do what needs to be done. 'Rather than thinking outside of the box, the Mittelstand has always been for enlarging it,' Andreas Woergoetter, an economist with the OECD, has said. 'They prefer incremental over radical innovation.'

The culture too is far from bean bags and ping pong tables. These guild-like enterprises are often led by founders with outsized personalities, who usually call the shots without the moderating influence of shareholders, partners, or even a board. 'It's just ego,' says Herbert. 'You have a lot of patriarchs who don't want to share what is great. But the people who work for them, they're sort of like a guy with a whip. It's a very brutal culture. Very often: no mistakes, no errors, perfectionism.'

And so there is a feeling building up that the old must give way to something new. 'I think what is novel now is that we are speaking about the New Mittelstand in a way,' says Herbert. In this way of thinking, the answer to Germany's waning relevance in tech is not to double down on the Mittelstand model that differentiates it from rivals in Silicon Valley and China, but to do exactly what the others are doing and build venture-backed, fast-growing startups that swiftly mature into public companies. 'There's so much more capital available which fuels the velocity of innovation getting into market,' Herbert continues. 'Venture capital and financial investors play an important role and financial instruments play an important role and we need to think in speed and scale and that's what is different from the past.'

It's not an either/or proposition. Many of the old guard have emerged as major investors in new startups. In late 2023, the Schwarz Group, which owns Lidl, and Bosch, Europe's largest car parts supplier, led a $500 million funding round for Aleph Alpha, the German rival to OpenAI. Much of this financing is in the form of a grant with no expectation of returns. The Schwarz Group has also poured $2 billion to set up an AI cluster in the state of Baden-Württemberg, expected to be the largest in Europe.

Susanne Klatten, the richest woman in Germany, who owns a fifth of the automaker BMW, sponsors UnternehmerTUM, a leading startup school in Munich. 'The goal is to have this ecosystem where you have established companies, family businesses and leading universities coming together,' Helmut Schönenberger, the institute's founder and CEO, told me. The school was the launchpad for some of Germany's most prominent new companies, like Celonis, the first German tech startup to cross a $10 billion valuation, and Flixbus, also valued at a billion plus, which operates bus services in over forty countries. A fifth of all venture capital raised in Germany flows to startups at UnternehmerTUM.

And so, even if the Mittelstand, still widely respected in Germany, is to fade in the face of a changing reality of what it means to do business in the modern world, it may yet have a lot to pass on to its successors. 'They benefit in a way from the culture of the Mittelstand,' says Herbert. 'If you think of manufacturing in the US, I would say the US is a very, very bad place for manufacturing, because you don't have the discipline, you don't have the culture and heritage. What China is doing is incredible, it's low cost. But the quality and the low cost of manufacturing in Germany, I think this is unmatched on this planet. It's still unmatched.'

7

Henrik Brandis started sailing when he was four years old. By the time he turned 45 he had won the Rolex Swan 45 World Championship, a major fixture in the international yachting calendar, four times. 'In competitive sailing you can choose pretty conservative strategies, which will probably hinder you to win really big: you don't lose big, but you don't win big, and on average it is not a successful strategy,' he says. 'I have the privilege to go sailing with professionals who want to win. I'm competitive as well. I like to win as well.'

Hendrik's competitive streak and penchant for high-risk, high-reward plays has held him in good stead in his day job as one of Germany's more prominent venture capitalists. His firm, Earlybird, was one of 107 new venture funds that came up in Germany during the dot-com boom of the late nineties. Twenty-five years later, only three are still standing, and Earlybird is the largest of them all.

Hendrik attributes his fund's longevity to a quality at odds with the maniacal focus that defines the Mittelstand: diversification. 'We managed to build up a truly diversified technology portfolio in days where everybody could only think of consumer internet,' he says, 'which helped a lot once the internet bubble started to burst in 2000.'

If there is at all anything like a new Mittelstand bubbling up in Germany, Hendrik is among a small group of people playing an outsized role in bringing it to life. Earlybird, as the name suggests, is an early stage investor, the largest of its kind in Europe. It invests in companies sometimes when they're only just a germ of an idea in a budding entrepreneur's mind. The firm wrote the first cheques that got some of the country's most ambitious new startups off the ground. Its portfolio includes Isar Aerospace, Germany's SpaceX, Aleph Alpha, the German OpenAI, and Lilium, the electric jet manufacturer.

Earlybird is plugging the gap for much needed American-style risk capital for early stage companies who traditional investors in the country's highly conservative business environment wouldn't go anywhere near. 'In Europe it's basically impossible to sell a pure dream, whereas in America that's what they expect from you,' says Dirk Radzinski, the founder of Xolo3d, a Berlin-based 3D printing startup. 'They expect you to sell what happens in twenty years and not what happens in two years. In Europe it's exactly the opposite.' He adds, 'The old German industry, they're super conservative, they're super happy if everything is developed, but they don't want to put anything to develop it, they wait until the last minute.'

It would be misplaced to explain away everything in terms of this all-encompassing cultural aversion to risk. Other factors are also at play. Structural differences in the way large-scale financial institutions like pension funds are organized on the two sides of the Atlantic also have a downstream impact on the amount of money available to finance innovation in these two regions.

'It's very simple,' says Johannes von Borries, the founder of UVC, a venture fund based in Munich. 'In the US the biggest investors in venture capital are the big pension funds, like CalPERS, the California state pension fund. They're so big that even if they invest 1 per cent or 2 per cent in venture capital it's a lot of money.' CalPERS, the California Public Employees' Retirement System, has almost half a trillion dollars in assets. In the first six months of 2023 alone, it poured $4.5 billion into venture funds, a fraction of its portfolio which nevertheless amounted to 15 per cent of all capital raised by US VC firms in that period. 'In Germany we don't have these kinds of pension funds. Why? Because in Germany when I make my pension payment it goes immediately out to pay for someone's pension, so there's no capital stock to invest, and this is a huge amount of money that's missing in Germany, and that's actually the biggest problem in terms of structure.'

And while thanks to firms like Earlybird, entrepreneurs now have access to six and seven-figure sums that makes it easier for them to get their ideas off the ground, they still find it hard to raise the hundreds of millions and billions that are needed to reach global scale. Aleph Alpha is the best-funded AI startup in Germany. But OpenAI has raised 30x more money. That capital advantage can be decisive.

Hendrik, an aerospace engineer by training, cut his teeth in the tech sector in the 1980s while working on the Eurofighter, one of Europe's most ambitious technology projects of the previous century. He has seen first hand that the continent punches on a global level when it comes to the quality of its technical minds. 'Technologies are both with respect to density and quality at least on par with the most innovative ecosystems of the world, namely the US and China,' Hendrik continues. 'Where we miss out in building global champions is in the growth phase. Either companies are sold to international players, or they simply remain underfunded and are therefore unable to pursue a really aggressive global growth strategy.'

Earlybird has tried to lead by example when it comes to

international growth. It opened an office in Palo Alto in the early 2000s to take on the marquee names in the US venture industry on home turf. The experiment largely failed. 'We totally misjudged and underestimated the challenge of getting access to the best deals in the US in a very well-funded market,' says Hendrik. Venture is a clubby business. It's not about what you know, it's about who you know. The best deals go to a small group of best-connected investors. It can be hard for outsiders to break into these in-groups. 'You have to be pretty lucky to find extremely good opportunities because so many people have looked at it before you and you must see something that nobody has seen yet.'

The firm's failed expansion to the US gave Hendrik the seed of an idea that would pave the way for the firm's future triumphs. If access to in-groups is what mostly drives performance in the venture industry, then the firm's strategy should be to expand not in geographies where these networks are already fully formed and thus hard to breach, but in places where they are yet to take shape. By building these networks in new markets, Earlybird too could benefit from the same advantages that older firms enjoyed in Palo Alto.

After its failed expansion to the US, the firm shifted its focus from the West to the East and opened an office in Istanbul to prospect for opportunities in Turkey and Eastern Europe. This proved fortuitous. In 2015, Earlybird's eastern fund made an investment in a small startup called DeskOver run by two entrepreneurs out of a shabby building in a modest part of Bucharest, Romania.

DeskOver, which makes robotic process automation software, would later change its name to UiPath. It debuted on the New York Stock Exchange in 2021 at a valuation of $36 billion, the largest IPO for a European-born company ever at the time. The event was seen as a veritable Cinderella story coming as it did not from the established European tech capitals of London, Stockholm or Paris but from the backwaters of Bucharest. The company's founder, Daniel Dines, came from humble origins. He lived long stretches of his life on a dollar a day and learned programming by reading secondhand manuals before he had the money to even buy a computer.

Earlybird's pivot from geographies heavily covered by the venture industry to those overlooked by it outperformed the firm's wildest expectations. Earlybird was the first outsider investor in UiPath. Its modest million-dollar investment would in six short years net the fund upwards of $2 billion, a return of 220,000 per cent, the greatest early stage venture bet in European history. It also established the firm as the go-to venture fund in an up-and-coming region. 'Earlybird is by far the market leader with respect to early stage investment in the Eastern European hemisphere,' says Hendrik.

I asked Hendrik if there was a bigger meaning to UiPath's success, that the longstanding gap in technological capabilities between the US and Europe, the source of much anguish on the continent, is now beginning to close. His response was measured. 'I would love to say it's narrowing; I think it's not true,' he said. 'My true perception is the US is growing, Europe is growing, but we are not growing at a faster pace. So I think the gap is like it has always been.' He adds, 'We're trying to build global champions here. Are we going to be successful in that? We'll see.'

8

The most pervasive image of Germany internationally is that it makes good things. The country has often been held up as something of a model for other industrialized nations as a place that was able to keep its manufacturing base intact even as globalization hollowed out other industrial boomtowns in the US and UK.

Industry accounts for almost a quarter of Germany's total economic output and employs up to a quarter of its labour force, more than twice as much as the UK. While US companies like

Apple, Google and Nvidia hardly make anything in the US any more; Mercedes, BMW, Audi and Volkswagen, the pride of German engineering, have managed to retain much of their production capacity at home.

But even the mighty German industrial juggernaut is not impervious to creeping de-industrialization. Its firms are increasingly under the dual pressure of digitalization from the US and low-cost competition from China. This is most visible in the car industry which has long been the most recognizable face of German industry abroad.

'Das Auto' is generally struggling to keep pace with more technologically savvy rivals in the US and China. Tesla is now worth more than all German car companies put together. BYD makes more EVs than all the German carmakers combined. The automotive sector's contribution to the country's exports has largely stagnated. Germany exported as many cars in 2022 as it did all the way back in 1990.

Much of German industry has also been flocking to China. In 2023, BASF, the chemicals manufacturer, whose founding dates to the *Gründerzeit* in 1865, announced that it would invest $10 billion in a state-of-the-art smart manufacturing plant that it claimed would be a gold standard for sustainable production, not in Bavaria or Hamburg, but in Zhanjiang in the Guangdong province of China. This came only months after the company shuttered a plant in its hometown of Ludwigshafen, which led to 2,600 workers losing their jobs.

'We are increasingly worried about our home market,' the company's CEO, Martin Brudermüller, told shareholders, noting that the company lost €130 million in Germany the previous year. 'Profitability is no longer anywhere near where it should be.'

How to stop this de-industrialization? Many are betting that more automation can stem the flow of investments and jobs abroad. Foreign labour is cheap. But robots are cheaper. And they can reduce the country's reliance on shaky supply chains and politically fraught relationships abroad.

Germany has a higher density of robots toiling away in its factories than any other Western nation. With 371 industrial robots for every 10,000 employees, it trails behind only Korea, Singapore and Japan as the most intensive user of robots, significantly ahead of the US and China. The government has been encouraging more automation as a part of its Industry 4.0 strategy to reshore production back to Germany.

And many of the country's newer generation of companies are riding this automation wave. Celonis, the first German tech startup to cross a $10 billion valuation, is a software developer that uses Robotic Process Automation, or RPA, to identify and automate repetitive tasks in businesses.

'The US is a little bit behind,' says Armin Schmidt, the founder of German Bionic, a robotics startup based in Augsburg.

German Bionic makes electrically powered exoskeletons, robotics suits that can be worn to enhance human strength. Mechanical outerwear that confers superhuman powers has been the stuff of science fiction for just about as long as the genre has been around. Dean Martin donned one in the spy flick *The Ambushers* in 1967 as did Sigourney Weaver in *Aliens* in 1986. 'The oldest patent I found was from 1892,' Schmidt tells me.

German Bionic's smart power suit, now in its sixth generation, won't exactly turn you into Tony Stark. But it's a practical tool to save your back. The device looks like a backpack with a hard shell which attaches to the shoulders, hips and thighs. It acts like a mechanical muscle that can enhance strength by up to 36 kg. The main application is to reduce the strain on workers who perform repetitive physical tasks all day: factory workers who lift heavy weights, baggage handlers at airports and nurses in hospitals. 'Our DNA is we want to empower humans; we don't want to replace humans because we don't believe that humans should be replaced,' says Schmidt.

This heritage of engineering excellence has made Germany one of the centres of the robotics revolution. But its hopes that the technology would act as a foolproof hedge against globalization haven't quite worked out as expected.

Not far from German Bionic's office in Augsburg are the headquarters of perhaps the most recognizable name in the global robotics industry. Kuka, founded in 1898, is one of the four biggest robot manufacturers in the world, no small achievement in an industry dominated by Asian players. Its distinctive orange robots are a familiar sight on factory floors everywhere; SpaceX, BMW, Hyundai, Volvo, all use Kuka robots in their assembly lines.

In 2016, Kuka was bought out by the Midea Group, a Shenzhen-based conglomerate, for €4.5 billion. It was hard not to see the irony. Kuka was seen as a symbol of Industry 4.0, the grand plan to out-compete foreign rivals via more automation. And then an overseas player went ahead and bought that same company. It used to be only German jobs that moved abroad, now it was entire companies.

The deal caused a political storm in Germany even as it was happening, but little could be done to stop it from going through. No European company was willing to match Midea's generous offer, and the transaction did not break any laws. The Kuka episode is now seen as something of a turning point in how Germany sees overseas acquisitions and regulators have since tightened the rules for foreign takeovers in strategic sectors.

The aftershocks of the Kuka acquisition still reverberate in Germany's tech circles almost a decade after it closed and featured prominently in my discussions in the country. I wasn't so much surprised by the opposition to it, I expected that to be the mainstream view, but I was surprised to find a strain of thinking that seemed supportive of the development. China is a big market, if they can't play here then we won't be able to play there, they're writing bigger cheques than anyone else, that sort of thing. But suffice it to say that it was a divisive topic, and the sentiment generally leaned more against than for.

But in the longer run that acquisition might not matter that much. 'In the future the robot manufacturers will be less important since their hardware is being commoditized more and more,' says Christian Piechnick, the founder and CEO of Wandelbots, a robotics company based in Dresden.

Robotics is going through something of a PC moment. There are about 4 million or so operational robots in the world, nearly all of them installed in industrial settings. Piechnick thinks that current adoption rates are less than 1 per cent of what they will eventually be, and numbers will inevitably rise into the billions. It will be the same hockey stick graph that we've seen with PCs, smartphones and the internet.

But in the same way that the original PC hardware manufacturers – IBM, Compaq, DEC – are long past their prime or no longer in business but the earliest software companies – Microsoft, Apple – are still thriving, Piechnick expects a similar dynamic to play out in robots. 'If you look at the stock of ABB for instance, over the past twenty years, it's basically a flat line,' he says.

Wandelbots makes robot agnostic operating systems. At present robots from different manufacturers run on their own proprietary software. Wandelbots is making something like Windows or Android for robots, which will make all robots interoperable and easy to use.

That's a tall ambition, to be the Windows of the robotics world. I asked Piechnick whether he felt the company was ideally positioned on the map to deliver on that objective. 'I would say it's the most demanding industry globally,' he said. 'And if you're able to fully make a customer happy here, you will make it anywhere on the planet.'

In Germany he has access to world-class technical talent and funding too had not been an issue. So how does he square his own cheery experience with the larger national discourse which is tinged with anxieties about Germany's uncertain positioning in the industries of the future?

'I think it's a bit of a German mindset to always complain and see the glass half empty rather than being half full,' he said. There are only a handful of companies in the world that have what it takes to be the next big thing in robotics software. And in that contest the fact that Wandelbots can call Germany its home is unequivocally a net positive. 'It's a race,' Piechnick told me, 'and I don't intend to be second place.'

CHAPTER EIGHT

Importing Genius

'It would be a shame if we invented this technology and then had to buy the applications of it back from others.'

I

The reason why we went from not hearing about AI at all to hearing about it all the time can in part be traced back to Canada. Some of the biggest research breakthroughs that power the contemporary boom in AI came from researchers working at Canadian institutions, with three scientists doing a disproportionate amount of the heavy lifting: Geoffrey Hinton at the University of Toronto, Yoshua Bengio at the University of Montreal and Rich Sutton at the University of Alberta.

The outsized role that Canada plays in AI research is the story of the difference that just a few individuals, in this case just three, can make in the technological relevance of an entire country and how those individuals don't even need to be from there to make it all happen.

Let's start with Geoffrey Hinton. To appreciate his contribution to AI it would be useful to maybe understand some of the big debates that have animated the discipline since the term 'artificial intelligence' was first introduced at a conference at Dartmouth in 1956, the event which marked its birth as a distinct field of study.

What is popularly referred to as 'artificial intelligence' is in fact an umbrella term that refers to a range of approaches that can be used to simulate human-like cognition in machines. Adherents of these divergent approaches can have an almost sectarian attachment to their preferred way of doing AI and passions aroused by these rifts have shaped this young discipline for the seven decades or so that it has been in existence.

Neural networks are one approach to AI that has been around since just about the beginning. Their basic premise is that artificial brains should operate on the same principles as human ones: a web of interconnected neurons that work in concert to process information. That sounds obvious enough except it's not. It's not a given that man-made systems should mimic their biological counterparts. After all, planes don't fly by flapping their wings.

Neural networks are probably best explained by comparing them to what they are not. Conventional computer programs run on explicit rules and predefined instructions that are hard-coded into these systems: complex versions of flowcharts that run the program down different paths based on actions taken by the user, if this happens then that happens, if this doesn't happen then the other thing happens. There is a branch of AI, called symbolic or rule-based AI, also known as Good Old Fashioned AI, or GOFAI, that takes this approach.

Say you're trying to teach a computer how to recognize a cat. The GOFAI approach would be to give it lots of rules about what makes a cat a cat: it has a tail, ears, its furry, and so on. Every time the program sees an object that ticks off everything on that checklist it knows that what it's looking at is a cat. Cracking the problem is all about giving the program the right set of rules.

Neural networks would take a different route to solve the same problem. In this approach the researcher would feed the computer millions of images of cats, all tagged 'cat', and then the system can go through these images to identify patterns and learn what makes a cat a cat. Instead of being given rules, the system is given examples, and it learns through pattern recognition.

This is what makes neural networks a lot like human brains. A child doesn't learn how to recognize a cat by learning rules about what defines catness but by looking at lots of examples of something that keeps being called a cat.

And these rules-based versus examples-based approaches can be used as two different methods to solve the same problems. So if a system has to be taught how to play chess, the GOFAI approach would be to feed it the rules of the game, the different positions on the board, and moves that the pieces can make. The neural nets approach would be to feed it lots of chess games so it can recognize patterns in gameplay and learn how to play for itself.

Both these approaches have been around since pretty much the beginning of AI as an academic discipline. As far back as 1957, researchers at Cornell built the perceptron, a five-ton computer the size of an entire room, which in the words of its inventor Frank Rosenblatt was the first machine 'capable of having an original idea'. That was a bit of an oversell as the machine could only perform simple tasks, like identifying cards marked on the left from those marked on the right, a modest but conclusive demonstration that machines could be taught to learn on their own.

The perceptron is now considered the world's first AI system. When it first launched this 'thinking machine' had the same innervating effect on the media that ChatGPT would have more than sixty years later. 'NEW NAVY DEVICE LEARNS BY DOING: Psychologist Shows Embryo of Computer Designed to Read and Grow Wiser', wrote the *New York Times*. 'Indeed, it strikes us as the first serious rival to the human brain ever devised,' observed the *New Yorker*.

The enthusiasm waned as the perceptron's limitations soon

became apparent. It could only perform the simplest of tasks and those too under rigid experimental conditions. Widespread scepticism soon followed and Marvin Minsky, an MIT professor who towered over AI as the discipline's most prominent figure of the previous century, went to the extent of writing an entire book called *Perceptrons* in 1969 dedicated to assailing their shortcomings. The following year Minsky won the Turing award, the highest award in computing.

Frank Rosenblatt's career took the opposite turn. With the fate of neural networks effectively sealed, his research interests drifted away from neural nets to the only marginally more dubious enterprise of injecting matter from trained rats' brains into untrained rats to see if that could also transfer learning. It was this that preoccupied his mind when he died in a freak boating accident in 1971 at the age of 43.

Neural networks fell out of favour as something of a borderline disreputable enterprise whose adherents were painted as occultists dabbling in voodoo and pagan ritual. It was not a pursuit for serious minds. Neural was a bad word and researchers went to great lengths to replace it in their papers with terms like 'functional approximation' and 'nonlinear regression'; anything that could dupe reviewers from seeing their research for what it was. From the 1970s all the way through to the early 2000s, bright young things doing their PhDs were told to redirect their energies elsewhere; neural nets were a dead end, the surest way to put a promising career behind you.

2

Geoffrey Hinton kept the faith even as the wider academic establishment shunned neural nets. 'In the eighties Geoff stood up and contradicted all of the United States' AI community, all of them, called them all wrong, and I knew a lot of them, and they were very smart people, so it was very bold of him to do that,' Garth Gibson, the founding CEO of the Vector Institute for Artificial Intelligence in Toronto, tells me.

Hinton was born in the dying days of the British Empire in the winter of 1947 in Wimbledon, England into a family of scientific royalty. His great grandfather was George Boole, the inventor of Boolean algebra which is the basis for all modern computers. His cousin Joan Hinton was a nuclear physicist who worked on the Manhattan Project. When he was a child Hinton's mother told him that 'he could either be an academic or a failure', and, for a while, it seemed like he was destined to be the latter.

When he enrolled as an undergraduate at King's College Cambridge in the late 1960s, Hinton couldn't figure out what, if anything, he was good at. He studied physics then chemistry then maths and then dropped out entirely only to reapply a year later to try his hand at architecture then switched to physics and then physiology before graduating with a degree in experimental psychology in 1970. After which he promptly proceeded to work as a carpenter in London, building shelves and hanging doors, just to make a living.

Hinton was persuaded to return to academia a couple of years later and he enrolled in a PhD programme in artificial intelligence at the University of Edinburgh. Many years later, a colleague would introduce him at a conference as someone who had failed at physics, dropped out of psychology, and then ended up in artificial intelligence, a field with no standards at all. Hinton likes to repeat the story with a caveat: 'I didn't fail at physics and drop out of psychology; I failed at psychology and dropped out of physics – which is far more reputable.'

It was at Edinburgh that Hinton gravitated towards neural nets. It was a dangerous flirtation. The early 1970s were the lowest point for the technique, a time when most researchers were getting out of this discipline rather than getting into it. But Hinton, still in his twenties, thought the obituaries were premature. 'Geoff's position was always that there is only one intelligent machine that any of us know of and it's the brain,' continues Gibson, 'that's the only one. Nothing else has been proven to be able to do this. And why would we expect that we can invent something from scratch that has a chance of competing?'

And for this conviction Hinton spent most of his career largely on the sidelines of the academic mainstream. When he finished his PhD, he struggled to land even job interviews for teaching positions in the UK. He was compelled to move to California to the University of California San Diego, or UCSD, where the intellectual climate was more accommodating of fringe ideas. He bounced back and forth between academic departments in the US and the UK for a decade before finding a home for himself in the computer science department at the University of Toronto in 1987 which is where he's been ever since.

In the mid-1980s, Hinton presented a paper at a conference of around twenty AI researchers who had gathered at an old French manor-style estate outside Boston which served as a retreat for academics from nearby MIT. In attendance was Marvin Minsky. Hinton handed out paper copies of a maths-laden treatise on what he called the Boltzmann Machine, a kind of neural network that overcame the flaws of the perceptron that Minsky had assailed in his book fifteen years earlier.

As Hinton gave his talk from the front of the room, Minsky took out the staple from his copy of the paper and laid out the pages one next to the other on the table in front of him. He listened without comment and when the lecture ended he got up and walked out without saying anything, leaving the paper behind. Hinton later gathered the loose pages from the desk and mailed them to Minsky's office at MIT along with a short note that read: 'You may have left these behind by accident.'

3

Widespread scepticism held back progress in neural nets for the following three decades, all the way up to 2012 when Hinton along with two of his graduate students entered ImageNet, a prestigious computer vision contest at Stanford which invited researchers to submit AI programs that could identify objects in images. Hinton's submission, known as AlexNet, which was based on neural nets, blew the competition out of the water, identifying more images with fewer errors than any competing approach ever. This was not an incremental improvement over what had come before but a substantial leap that practically revolutionized the field of computer vision.

The 2012 ImageNet results were a seismic event in machine learning and much of the contemporary boom in AI can be traced back to that moment. It brought neural nets in from the cold, taking them from the margins of AI to making them its main event almost overnight. Neural nets are now seen as not just a viable approach to build machine learning models but, in most instances, the most promising method. Once derided as hocus pocus propagated by woo woo people, the technique is now practically synonymous with AI; these days whenever you hear the word 'artificial intelligence' it is likely it's a rough substitution for the more precise term 'deep learning by neural networks' to the exclusion of all other methods.

Neural nets are the backbone of some of the most high-profile examples of AI in use today, from ChatGPT to Facebook image recognition to Tesla's autonomous cars. Hinton, long seen as something of an oddball on the periphery of the AI movement, is now considered among the field's biggest if not *the* biggest star. In 2018, Hinton, along with two of his long-time collaborators, Yoshua Bengio and Yann LeCun, won the Turing award, the million-dollar prize that is considered computing's highest honour, the same accolade that Marvin Minsky, the man who single-handedly sank the reputation of neural nets, won nearly half a century earlier.

Hinton's story of overcoming formidable technical challenges and staring down entrenched attitudes of the scientific establishment is punctuated by enormous personal tragedies. These battles were fought while he lost two wives to cancer and himself suffered from a lifelong condition which makes it impossible for him to sit down for more than a few minutes at a stretch for the fear of slipping a disc. Hinton is forced to spend his waking hours on his feet or on his back, cannot fly commercial, and often sees students while lying down flat on his desk. He addresses his inability to sit with characteristic humour and often quips that he's afflicted with a 'longstanding problem'.

Take this tale of a scientist who stands, quite literally, in the face of orthodoxy and spends the better part of his working life in the wilderness only to be vindicated in the end when he effects a paradigm shift in the order of things all while braving enormous personal hardship, and you have an almost made for Hollywood story of what we somewhere in our collective consciousness imagine the scientific enterprise at its finest to be all about: man against the world, man wins; light against darkness, light wins.

I asked Hinton what made him stay the course through all these years even as he had so much arrayed against him. He said it was an innate trait that made him impervious to peer pressure; an openness to the possibility that everyone could be wrong. He recounts an incident from school when he was seven years old when the teacher asked the class: 'Where do good all things come from?' Hinton, the child of card-carrying communists, shouted 'Russia!'

When Hinton first moved to the University of Toronto in the 1980s, the school had a computer science programme that was well-regarded but not generally considered to be top-tier. And while he may have spent much of his life out of the mainstream, the mainstream has now come to him. ImageNet sparked a war for AI talent coming out of Hinton's lab in Toronto. In 2013, Google paid $44 million to acquire DNNresearch, a startup built around Hinton's research, which had only three employees, no products and no financials. Hinton and Alex Krizhevsky, his collaborator on AlexNet after whom the project was named, went to work for Google. The other collaborator on that landmark paper, Ilya Sutskever, is now one of the most recognizable faces in AI as one of the co-founders and Chief Scientist of OpenAI who was ousted from the company after a failed putsch.

Now in his late seventies, Hinton still maintains an active research agenda though he has taken a step back from advising students. 'Supervising research means you have to be able to peer into your students' minds and see how they are thinking, and I've been finding it harder to do that lately,' he tells me. Hinton's former advisees and collaborators are in high demand wherever AI is done in the world and can be found in the upper reaches of just about all the AI divisions of all the major tech companies from the US through Europe to Asia.

4

The fact that the University of Toronto could produce the biggest breakthrough in AI in a generation and its students are hot properties in the most advanced divisions of the most advanced tech companies in the world is tremendous validation for Canada's ability to punch at the highest level in frontier computing. But it also raises uncomfortable questions.

Why is the country training the brightest minds in artificial intelligence in publicly funded universities only to see them leave to go work for American, Chinese and British companies? Why were big scientific breakthroughs in machine learning happening in Canada but its commercial gains accruing to overseas entities?

Why is Canada so good at AI research but so bad at cashing in on the technology's potentially transformative economic potential?

'It was basically invented here but the economic value was accruing to West Coast tech companies and Chinese tech companies,' says Jordan Jacobs, the Managing Partner of Radical Ventures, a venture capital firm based in Toronto. 'And it would be a shame if we invented this technology and then had to buy the applications of it back from others.'

Jordan is a major figure on the AI scene in Canada. In 2017, he was among eight people who founded the Vector Institute for Artificial Intelligence in Toronto. The idea was to take advantage of Geoffrey Hinton's presence in the city by building an institution around him that can take his legacy forward. Hinton is one of the founders and Chief Scientific Advisor to the institute.

Typically, when countries build up their AI capabilities, they pour funds into university departments and research labs. Vector is unique in that it's a new institutional form that sits on top of the pre-existing university structure. It pools resources from over twenty-four universities and acts as a bridge that connects departments across institutions.

At its core, Vector is a research institute which aims to compete with the likes of Harvard, Stanford and MIT in AI. It does this by offering grants and scholarships and by bringing promising talent to Canada. But the other part of its mandate is to apply this research to real world applications. The institute is responsible for improving the adoption of AI in existing industries and helping startups that are creating new ones. It does this by incubating companies, organizing workshops and putting AI talent in touch with companies that need it.

But the ambition of Vector goes beyond just creating more AI companies. It wants to be a catalyst for transforming the very character of the Canadian economy. Canada is unusual among advanced economies in that it still relies disproportionately on raw materials, mostly oil and trees. It's the world's second largest exporter of wood and fourth largest exporter of oil. Both have an ending that's not that far away and are increasingly in the crosshairs of environmental activists. The grand plan is to swap out raw materials for AI-powered companies in the country's economic base.

'We can't sit there and say well we'll devalue our way to be competitive; the idea that we export raw materials and live off that,' says Ed Clark, the chair of the Vector Institute. 'It's obvious that oil is not going to carry us forever, so we've got to become better at raising these artificial intelligence firms than anyone in the world.'

Clark, now in his late seventies, is seen as something of an elder statesman of Canadian business. But he took an unusual path to get there. Born into an academic family, his father was the founder of the Department of Sociology at the University of Toronto, and his mother studied economics at Columbia before World War II, a rare feat for a woman at the time. Clark, like his brother, got his PhD in economics at Harvard in the 1970s. He joined the Canadian government soon after where he served in high-profile positions for a decade. But then what he thought would be a long career in the civil service was unceremoniously cut short in 1985 when the country's then prime minister, Brian Mulroney, personally fired him.

'He called every single Canadian institution to which I applied and said: you go to war with the government of Canada if you hire him,' Ed tells me. 'I would walk down the street in Ottawa and people would literally cross over so they wouldn't run into me.'

And so Clark, at 37 years of age, with four kids and no money, was forced to leave a promising career in government and his country to start a new life in the US where Merrill Lynch was the only company willing to hire him. When the Canadian prime minister's ultimatum reached his new employer the company's president told Ed: 'Nobody tells Merrill Lynch who we can and cannot hire. You're hired!'

The middle-aged Clark went from the top of the food chain in the Canadian government to the bottom of the pyramid in corporate America. 'I would walk around and carry the briefcases of 25-year-old investment bankers who would say: read that document, check it for typos,' he recalls. He would quickly make up for lost time to build a wildly successful career in investment banking. He retired as the CEO of TD Bank where over a span of twelve years he transformed a once middling enterprise into the second largest company in Canada and one of the ten largest retail banks in the US, cementing his reputation along the way as one of the top businessmen in Canada. 'My wife always says Brian Mulroney made me a multimillionaire because I probably never would've quit the government and gone down an entrepreneurial route,' he says.

Clark, who was made an Officer of the Order of Canada in 2010 and inducted into the Canadian Business Hall of Fame in 2016, describes himself as a Canadian nationalist, and would like to see his country produce more globally relevant companies. 'If I have a passion in addition to my philanthropic side, it is: how do we create great Canadian companies, not just great Canadian subsidiaries of international companies,' he says.

When Clark talks, people listen. When Jordan Jacobs was looking for someone who could be the chair at Vector, he thought that he needed to look no further than Clark. Ed had the stature and relationships to get the right people behind the initiative and the entrepreneurial drive to get things moving quickly. 'Ed was a very critical catalyst to the whole thing,' Jordan tells me. 'He had the relationships where he could call the CEO of a big bank and say, you're putting in five million, hang up the phone, and then say to me, go tell him why he's putting in five million.'

Clark, who retired from his banking career over a decade ago, has found new purpose in his role as the country's AI whisperer in chief. 'I do actually believe in the case of AI; if we don't win all our conventional industries will fall behind,' he says. 'It's existential. We have to win this battle.'

5

Vector was one of three institutes set up as a part of the Pan-Canadian AI Strategy – the world's first artificial intelligence strategy. The other two are the Montreal Institute for Learning Algorithms, or Mila, in Montreal, and Alberta Machine Intelligence Institute, or Amii, in Edmonton. Canada takes egalitarianism seriously and so when it came to AI it wanted to spread the peanut butter evenly across its major provinces.

If Vector is closely affiliated with Geoffrey Hinton, Mila is built around Yoshua Bengio, a professor of computer science at the University of Montreal. Yoshua worked closely with Hinton in the early years when neural nets were still considered a dead end so it was only fitting that the two shared the Turing award in 2018. Without their partnership, the contemporary boom in AI may not have happened.

'Geoff, Yann and I were pretty close,' Bengio tells me. 'We still talk to each other. But I think it was more crucial in those years when deep learning could have not happened. I was under a lot of pressure to work on other things, because my students wanted to have a job after they finish.'

Bengio is the most cited computer scientist in the world and could practically walk into any CS department anywhere in the world and land a professorship. He has also fielded many offers from big tech, including a very serious overture from Microsoft. But even as his Turing co-recipients have spent long stretches at big companies – Hinton worked at Google and LeCun is the head of AI research at Meta – Bengio has stayed firmly within academia and that too with the same institution that he's been affiliated with since 1993, the University of Montreal. I asked him why.

'There's something about the, let's say, my attachment simply to the country that allowed me to be who I am,' he said. 'I felt like going to the US would be probably very good for my career but I wouldn't feel good about myself. I benefited from the education system here which is essentially public and then the other thing is I had this crazy vision that we could create something like Silicon Valley here in Canada.'

Mila is the institutional vehicle bringing that vision to life. Its goal is to assemble a critical mass of AI researchers in Montreal who can come together to create the network effects that make places like Silicon Valley thrive. The institute is primarily a partnership between the University of Montreal and McGill and is tasked with producing world-class research and cutting-edge startups in machine learning.

Bengio credits the Canadian Institute for Advanced Research (CIFAR) for providing the initial funding for neural nets when the discipline was still unpopular in the academic mainstream. CIFAR is also the institute that developed and implements the Pan-Canadian AI Strategy under which the three major AI institutes – Vector, Mila and Amii – were established. 'It invests in long term bets and it invested in our deep learning dream,' says Yoshua. 'It was not a lot of money but it gave us, if you want, the moral support and the intellectual stimulation.'

CIFAR supports long-term, global, interdisciplinary collaboration in frontier science, everything from understanding consciousness to making quantum materials. The institute refers to itself as a 'network of extraordinary minds'. It is seen as being among the more risk-tolerant agencies that funnel public funds into basic research, sometimes making generational bets on promising ideas.

CIFAR has a relatively modest annual budget of only around $30 million compared to what's available stateside like DARPA's hefty $4 billion balance sheet. But it has amassed a solid track record of making directive bets on high-risk, high reward plays: a useful model, Bengio says, for other countries interested in funding bold ideas in a resource constrained environment.

When the Pan-Canadian AI Strategy was launched in 2017, the Canadian government allocated CAD $125 million for the three AI institutes – Vector, Amii and Mila – for their first five years. In 2021, it allocated another CAD $443.8 million for the next ten years. That sounds like a lot but it's not. Canada's overall public spend on AI research is dwarfed in comparison to what the competition is spending on shoring up AI research.

In 2023, the UK government pledged a billion pounds for AI research. In 2021, the US National Security Commission on Artificial Intelligence chaired by Eric Schmidt proposed to double the country's annual spend on non-defence AI R&D to $32 billion. Precise figures for Chinese non-military investments in AI are hard to come by but American estimates put them on par with US public spending.

'We're never going to outspend the UK, China or the US. We don't even come close,' Valerie Pisano, President and CEO of Mila, tells me. 'So if we're not going be bigger, if we're not going to be richer, then what we need is to be faster, we need to be smarter, we need to be more creative.'

Canada knows it can't compete using the same playbook as countries with more substantial resources. So it has tasked Vector, Mila and Amii to come up with different strategies that don't simply mimic what's happening elsewhere. 'The idea is we need to go about this differently and we need to be bold,' Pisano continues. 'The typical approach, give money to small groups within universities, is probably not agile enough, not aggressive enough, not creative enough, not big enough given this is not a game that is regional or national, it's international.'

When it comes to winning at AI, Canada cannot make it about money multiplied by effort, it will have to do something different. 'We use this term in Mila: it's innovating in innovation. It's not just about being in innovation, it's about being able to pilot new models, break, come back, try differently, do partnerships and do it hyper quickly,' says Pisano.

One strategy Mila is betting on is to make cutting-edge research from academia rapidly accessible to startup companies so they can build it into their products before anyone else in the world. The presence of Yoshua and Mila in Montreal is a huge boon for startups trying to bring new AI products to market faster than overseas competitors.

BrainBox AI is a Montreal-based startup which uses artificial intelligence to optimize energy usage in commercial buildings. The company uses deep learning to collect various data-points including weather, occupancy and thermal conditions to optimize energy consumption in real-time creating significant gains in energy efficiency.

The company's CEO, Sam Ramadori, tells me that being a part of Mila helps the company feed on talent and ideas coming out of one of the most advanced AI ecosystems in the world. The close feedback loop between AI researchers developing new techniques and companies like his that are using them in real world applications helps startups build, test, learn and iterate in faster and faster development cycles.

'It's a good healthy ecosystem that we're in the middle of and I think it matters, I think it matters a lot,' says Ramadori. 'Especially when you have a new technology where you have nowhere to turn. If I was coding traditional software there's nothing new to that. You can even get a team in some nation far away to code that whole thing for you. We're not there with AI. So you need that bouncing back and forth to be testing the latest and greatest.'

6

The third sister institute to Vector and Mila is the Alberta Machine Intelligence Institute, or Amii, in Edmonton, the capital of Alberta in Western Canada. Like Geoffrey Hinton at Vector, and Yoshua Bengio at Mila, the major figure associated with Amii is Richard Sutton, who is a Chief Scientific Advisor to the institute and a professor at the nearby University of Alberta.

Sutton is the world's foremost authority on Reinforcement Learning, or RL, a third approach to artificial intelligence different from neural nets and symbolic AI discussed earlier which in recent years has taken off as a popular technique to build more powerful AI systems.

RL also takes inspiration from animal models of intelligence, specifically behavioural psychology. RL systems learn through trial and error to adapt to new and complex circumstances. An RL program interacts with its surroundings by carrying out actions and receiving feedback in the form of rewards and penalties. Over time, the program learns to do more of the actions for which it receives rewards and less of the actions for which it receives penalties.

So if a symbolic system would teach a computer how to recognize a cat by feeding it rules for what makes a cat a cat, and neural networks by showing it lots of examples of a cat, RL systems would take the approach of feeding the program lots of images and giving it a reward every time it correctly identifies a cat and a penalty every time it doesn't. The algorithm initially gives random answers to prompts but as its cat-detecting behaviour gets reinforced it learns to give more accurate answers.

Like neural networks, reinforcement learning too has been

around since the beginning of the discipline with some of its earliest mentions coming up in the work of Alan Turing as far back as the 1950s. But until recently it was not seen as a particularly promising approach in the AI community with none other than Geoffrey Hinton being among the sceptics.

Like Nobel Prize winners, Turing recipients are invited to give a lecture about their life's work. Hinton gave his lecture in the summer of 2019 in Phoenix, Arizona during which he took a not too subtle dig at reinforcement learning: 'There are two kinds of learning algorithms, actually three, but the third doesn't work very well,' he said. 'That is called reinforcement learning.' As the crowd burst out in knowing laughter, Hinton went further: 'There is a wonderful *reductio ad absurdum* of reinforcement learning,' he added, 'it's called DeepMind.'

It was Hinton's turn to play Minsky, the establishment figure who dismisses the other thing just for the fuck of it, only to have to change his mind later. Reinforcement learning has now become de rigueur to newer and more powerful AI systems, not least because of work done at DeepMind. RL is central to AlphaZero, the DeepMind bot which can beat any human at Go, shogi, and chess; a vast improvement on its previous version AlphaGo, which AlphaZero can also beat.

One of the main differences between AlphaGo and AlphaZero is that the former is trained on examples of games like neural nets are, whereas the latter learnt solely through self-play reinforcement learning. And this is why AlphaZero's abilities are more general, it can play multiple games at superhuman level because its training is not based on learning from examples of previous games.

Reinforcement learning is also used by Covariant, a California-based company which makes AI-powered robots which are used in warehouses to automate processes like picking, sorting and assembling items. Hinton, who has a dim view of religion, experienced the closest thing to a come to Jesus moment when he decided to become an investor in Covariant, tweeting later: 'I had made a rather small investment (I don't want to reinforce reinforcement learning) but now wished I had invested 100X more.'

The one man most responsible for bringing reinforcement learning from the sidelines into the mainstream is Richard Sutton, who practically wrote the standard textbook in the field which has now been in print for over twenty-five years. Sutton, who with his long scraggly beard and unvarnished manner looks more like an archetypal philosopher than computer scientist, was born in Ohio and studied psychology as an undergraduate at Stanford before receiving his PhD in computer science at the University of Massachusetts Amherst in 1984. His career followed a relatively standard trajectory until 2003 when, while still in his forties and working for the AT&T Bell Labs in New Jersey, he was diagnosed with cancer which left him wondering if there was any point in doing anything any more.

'I thought I was going to die soon, it was all kind of surreal,' Sutton tells me. 'I just got tired of hanging around waiting to die so I thought I'd take a job in Canada.'

Sutton wasn't really thinking about making optimal career moves when he took up a teaching position at the University of Alberta in the winter of 2003. When he stood in front of his first class he told his students that he might not be around to finish the course. He had endured four major surgeries, chemotherapy and immunotherapy after an aggressive melanoma had spread to his major organs and brain. The university, which knew of his health struggles, had taken a gamble in hiring a well-regarded but ailing faculty member from abroad. It worked out brilliantly.

Turns out that the man who thought he wouldn't last the fall semester still had his best years ahead of him. Sutton would survive his five-year battle with cancer and two decades later, now in his mid-sixties, an age when most in his peer group have already gone into retirement, he is still busy doing his best work.

Sutton is considered among the dozen or so top minds in all of computer science, the sort of name that is whispered about every time it's Turing season. Along the way he has elevated the fortunes of the University of Alberta's computer science programme, previously an obscure research community on the edge of the sub-Arctic, into what is now considered a premier place to study machine learning. 'The pride is that the best place to study reinforcement learning in the world is the University of Alberta,' he says.

So what changed that made reinforcement learning a preferred option for building more advanced AI systems? The answer, in no small part, lies in the availability of more processing power. The objection that sceptics like Hinton used to make was that RL is too inefficient. If you have to teach a program how to play chess by trial and error then it just takes too many simulated attempts to get there; as opposed to showing it examples of previous games or hard-coding the rules. The dramatic expansions in compute power have blown past that problem with brute force. It is now possible for algorithms to be painfully inefficient and yet still startlingly effective.

Take for instance OpenAI Five, the program that plays Dota-2 at a world champion level, which was built on reinforcement learning principles. It took the agent 45,000 simulated years, or 250 years of gameplay per day, to learn to play. The program consumed 800 petaflop/s-days of compute over ten months of training. It would take an average laptop to run continuously at maximum capacity for over 200 years to take in the same amount of compute. Training OpenAI Five took 1.1 gigawatt-hours, or about ninety-two years of electricity use for an average US home. Proponents of RL like Sutton have long maintained that the technique always worked in principle but needed more powerful computers to emerge before their utility could be demonstrated.

'You have to appreciate that the increasing compute power makes it possible,' says Sutton. 'It's necessary but it's not sufficient. The bigger lesson is that as increasing compute power becomes available, the most powerful methods will be those that scale with compute power. But you still have to find those methods. The compute power itself doesn't do it for you.'

Sutton thinks that reinforcement learning is going to be more suited to building more powerful AI systems than the rival methods. 'I don't consider deep learning to be a good solution,' he says, and shares that he isn't much impressed by large language models like ChatGPT; to equate what they do with intelligence would be to diminish the meaning of the term. 'I take intelligence seriously,' says Sutton, 'it's the most powerful phenomenon in the universe,' adding that, 'OpenAI is mimicking people, it's not intelligence. It's not the ability to achieve goals. It's not a powerful thing to mimic people; it's going to be a popular thing but it's not going to be a powerful thing.'

It's not just OpenAI. Sutton is underwhelmed by much of what passes for progress in his field across the border in the US. 'Who in the US is trying to figure out how intelligence works?' he asks. 'You go to the universities and they're fixating on applications rather than understanding the mind, and I wish the places were better.'

Sutton thinks that Canada punches above its weight in basic research in large part because the country's main government body for funding research, the Canadian Natural Sciences and Engineering Research Council, or NSERC, operates very differently from US agencies like the National Science Foundation. 'Everyone gets a small grant and you get to say what you really want to do so you don't have to make up a story that's not true. It's smaller and more focused.'

In Sutton's view there are philosophical differences in how AI is approached in the US versus his research community in Canada. In the US the bent is more commercial, the frontier of AI research lies with companies trying to figure out practical applications. Whereas in Alberta the inclination is more towards a fundamental intellectual enquiry into the workings of the mind.

'I imagine myself as a part of a longer tradition of research trying to understand the mind and there are these big questions and we don't know how to answer them,' he says. 'And the other mindset is, no, we've got to get things to work somehow, and we won't really understand it so much, but it will work, it'll do something, it'll be really cool and impress everybody, the goal is to generate excitement rather than answer the longstanding decade- or century-old questions.'

In 2022, Sutton unveiled the 'Alberta Plan for AI Research', a twelve-step plan to place the research community in Edmonton at the forefront of solving the puzzle of intelligence. He believes that there is a one in four chance that human level artificial intelligence will emerge within this decade and that his research group will be the one to do it. 'Understanding intelligence is a grand scientific prize that may soon be within reach,' he told a packed auditorium at a talk organized by Amii in the summer of 2022. 'The Alberta Plan is a direct run at this great prize.'

'It's hard to say how big it is, it's comparable to the rise of life on this planet,' he said. 'It's when life figures itself out and makes more life not by replication but through design.'

7

There is broad agreement that Canada owes its place of prominence in AI largely to the presence of three people: Geoffrey Hinton, Yoshua Bengio and Rich Sutton. First came the three scientists, then came their graduate students, then came breakthroughs like AlexNet, then came the big tech companies looking for talent, then the big institutes and high-value companies, and then, finally, the big money. But first came the three scientists.

In one way this shows the power of individuals to change the relevance of an entire country to a key area of technology. But seen another way, this is not just the story of the triumph of individuals but of policies that manifested them. If Canada has won the AI lottery three times over, it's in large part because it has spent years buying up all the lottery tickets. The rise of its tech sector is not the product of good fortune but sound public policy which for decades had been engaged in a bold social experiment to find the finest minds the world has to offer and lure them to Canada.

All three AI pioneers are first-generation immigrants: Geoffrey Hinton was born in the UK, Yoshua Bengio in France, and Rich Sutton in the US. They are the norm and not the exception; most of Canada's tech headliners were also imported from abroad. Alex Krizhevsky, Hinton's collaborator on AlexNet, was born in Ukraine. The other co-author, Ilya Sutskever, was born in the former Soviet Union.

Tobias Lütke, the most celebrated entrepreneur in the country, who founded Shopify, which, at a market value of $200 billion, is the most valuable company to come out of Canada in the twenty-first century, was born in Germany. Christian Weedbrook, the founder of Xanadu, a prominent quantum computing company, was born in Australia. Roham Gharegozlou, founder of Dapper Labs, the most valuable privately held new tech company in Canada, was born in Iran.

Their arrival in Canada was not an accident but the outcome of a deliberate attempt by successive Canadian governments to populate this vast but mostly empty country with overseas talent.

Canada is the world's second largest country by landmass. It has almost as much land as all of Europe. But it has only about 40 million people, roughly the same as California. Canada's population density is among the lowest in the world, with only about four people for every square kilometre. Compare that to Bangladesh which has 1,300 people for every kilometre, or Singapore which has 8,000. And this native-born Canadian population is shrinking further: the country's fertility rate is below replacement, nearly half the global average.

More immigration has long been seen as the answer to this demographic deficit. As far back as the 1960s, the country was putting in place the groundwork for what is today a still unique policy regime among countries that are taking in the world's migrants. While the purpose of the immigration bureaucracy in most countries is to keep people out, in Canada it is to bring them in, with the administration setting ambitious targets for the number of immigrants it admits every year.

Canada gets roughly the same number of legal immigrants as the US, about 450,000 people a year, even though it only has a tenth of its population. Almost a quarter of the country's population was born overseas and it is projected that within twenty years a third of all Canadians will have immigrant backgrounds. Already, more than half of the residents in cities like Toronto, Mississauga and Brampton are foreign born, a much higher proportion than just about any other city in the world.

Some advocates propose numbers that are orders of magnitude above current levels. In 2014, Dominic Barton, the then head of McKinsey, launched a lobbying group which would later be known as the Century Initiative, a bold but controversial plan to increase Canada's population to 100 million by the end of the century. The project, backed by other corporate heavyweights associated with places like BlackRock, envisioned the formation of mega regions not unlike those found in Asia and identified 'a genius for getting along' as an area of exploration.

'We have an immigration strategy that says if all the smart people in the world want to come and live in Canada, why don't we have them come and live in Canada?' says Ed Clark. 'We get 400,000 immigrants a year, 325,000 of them are highly educated people that want to come here. What's not to like about them?'

That sounds like a sensible strategy and one would imagine that it would be the norm among countries that are popular destinations for migrants. It's not. Canada is a radical outlier when it comes to openness to outsiders and the country has maintained broad public support for immigration at a time when most countries have been closing their doors.

It would be easy to conclude that this is simply a matter of economic necessity: a shrinking working population and an end to a natural resource base that's not that far out means that the country has no option but to fill the deficit with more people from abroad. Immigration already accounts for a full 100 per cent of the growth in Canada's labour force. But this economic take would be a simple way to look at a complex matter. Much of the rich world has been depopulating for years and yet there is little enthusiasm for more migrants practically anywhere other than Canada.

Germany's native-born population has also been shrinking at about the same rate as Canada's and it takes in roughly the same number of immigrants every year, which translates to a smaller proportion in relative terms, since Germany has twice as many people as Canada. And yet the two countries could not be more different in how this wave of migrants is received.

None of Canada's five parties represented in parliament have an explicitly anti-immigrant agenda. Pierre Poilievre, the leader of the Conservatives, the country's second largest party, has pulled his party further to the right and is sometimes compared to Donald Trump in his strident populism. And yet even he has publicly insisted that 'the Conservative party is pro-immigration' and is himself married to an immigrant from Venezuela.

In Germany, divisions over migration have become the headline theme in the country's politics which have manifested most vividly in the rise of the far-right Alternative for Germany (AfD) party, which went from being an extremist fringe group with no representation in parliament to becoming the second strongest party in the country, all within the span of a decade.

As Germany goes, so goes the rest of Europe. Economic arguments are a hard sell when the objections are mostly cultural, that the huge influx of foreigners is diluting the attributes that make the people who they are. For immigration sceptics the primary question that needs to be asked is not how much of our own prosperity are we willing to share with outsiders, or, conversely, how rich can we get off these new people, but who are we?

In the Nordics, I heard concerns about how immigration was making the population less blonde. In Denmark, I heard a woman ask out loud whether the immigrants coming into the country could ever fully understand 'what it feels like to be a Dane', and as the country takes in more and more people who are less and less like themselves, if the notion of what it means to be Danish is itself at risk.

It's not a sentiment that's found only in the rich world. China's population is shrinking faster than that of Europe thanks to demographic carnage wrought by the one-child policy which economists predict is likely to have a knock-on effect on the country's growth prospects. And yet there is little, if any, political support for more outsiders.

China has 1.4 billion people out of which only a million, or less than 0.1 per cent, are immigrants. Compare that to 15 per cent in the US, 19 per cent in Germany, and 30 per cent in Australia. Even North Korea has a higher percentage of immigrants than China. In 2017, China's leader Xi Jinping told the then American president Donald Trump: 'We people are the original people, black hair, yellow skin, inherited onwards. We call ourselves the descendants of the dragon.'

So the question isn't why are the others not more like Canada but why is Canada not more like the others? Its demographic character is changing as much if not more than most other places. Canada has the smallest native-born population of any of the G7 nations and yet is taking more migrants relative to its population than anyone else in the G7. And these migrants are as different culturally as those being turned away from other immigration hot spots, a large proportion of whom are coming from Asia.

Where do Canada's migrants come from?

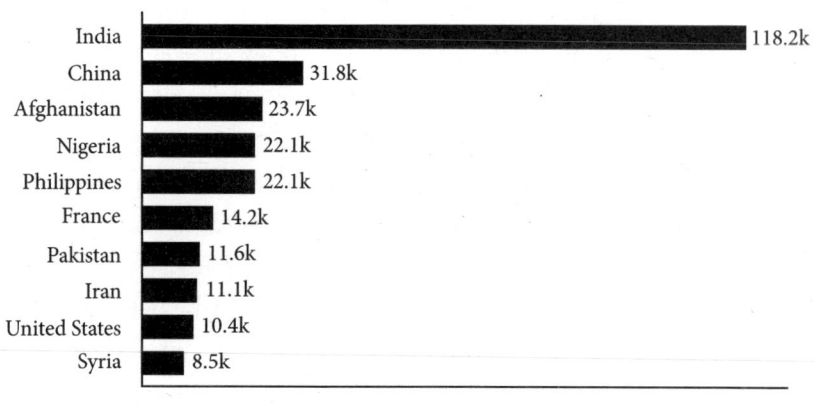

Source: Council on Foreign Relations (2024)

A big part of the answer is that while Germans and Chinese might have a fixed sense of what it means to be them, it's much harder to pin down what it means to be Canadian. People who carry that label seem to understand that they embody a fluid concept that exists in a state of permanent revision. It's an elastic identity. While other cultures anguish about how they can hold on to their originalness, Canadians seem curious to find out what their remixed version would look like, not resisting but leaning into faster iterations of what it means to be them.

This open-mindedness has made Canada the preferred option for immigrants who think they might have a harder time getting accepted elsewhere. Geoffrey Hinton's decision to take up a teaching position in Toronto was motivated in part by the fact that he felt that his two children, who are adopted from Peru and Guatemala, were more welcome there.

Ed Clark tells me with obvious pride that three of his grandchildren are half-Chinese and recounts an incident from a meeting at the Brookings Institute, where he is on the Board of Trustees. 'I sat in a Brookings thing and they had a black woman talking about AI and said: "Well let's be clear here, this is a white male business." And I turned to the guy beside me and said, should I show you a picture of the Vector researchers? I'm sure we've got a white male. I'm sure, I think, I can find one. But we've got Iranian women, we've got Russian women. And so you look at that and you say – that's our future.'

It would be erroneous to conclude that Canadian attitudes to immigration can be explained entirely by an intrinsic openness to outsiders. Geography plays a big role. The country has the benefit of isolation: the world's richest country underneath, endless ocean to the left and the right, and a block of ice up north.

Migrants to the US and Europe can cross the border on foot and millions have done just that. Those who want to move to Canada usually have to at least be able to buy a plane ticket. And that rudimentary filter produces a certain kind of immigrant: educated, middle-income and lawful. Other countries get immigrants that they can't turn away. Canada gets the immigrants it wants. And that makes all the difference.

'I literally studied what is coming out from the studies on diversity beyond the fact we all think it's the right thing to do or not. Does diversity really lead to greater creativity? Company performance? And the studies are unanimous,' says Valerie Pisano, the CEO of Mila. 'There's not a single study in the world that has contradicted the statement that greater diversity leads to greater performance. And I think Canada has that as a benefit. And certainly the tech and startup space is one of the places where you see that very vibrantly.'

It remains to be seen whether this enthusiasm for immigration can be sustained. In recent years there has been a significant revision in attitudes towards newcomers in Canada. While in 2022 there were only 27 per cent of Canadians who felt that there were too many immigrants entering the country, by 2024 this figure had surged to 58 per cent, representing the highest level of concern about immigration since 1998. In response to this changing public sentiment, the government has significantly reduced immigration targets and it's still an open question how these measures will impact the country's tech sector.

8

Canada has been upping the ante in the competition to attract the most industrious migrants by making it easier for them to move there. The country's immigration system is held up as something of a model and frequent comparisons are made to the comparatively arduous process that immigrants are subjected to when moving to other destinations like the US.

The US work visa, known as the H1B, essentially works like a lottery. In 2023, only about one in nine applicants got an authorization to work in the US. In Canada these visas are uncapped, and the country will take in as many workers as can be absorbed by its workforce. The US system doesn't differentiate for quality, an applicant is an applicant, and less qualified candidates are frequently chosen over more qualified ones. In Canada the quality of the applicant is at the centre of its points-based system which evaluates and ranks candidates by measures which include factors like age, language skills and educational background.

But the most appealing aspect of Canadian immigration is its efficiency and predictability. While even the lucky ones who get their H1B approved in the US still must jump through hoops and spend months in limbo to get their applications processed, in Canada the paperwork can be sorted in a matter of weeks.

'If I'm competing with Apple for someone from Switzerland or from the UK, Apple says to them: I can put you in the H1B visa lottery in six months and you may or may not get the visa; I can say to them: you can start two weeks from Monday,' says Jordan Jacobs. 'Almost everybody wants to have some definitive understanding of where their life is going to be, so that creates a giant advantage.'

Canada has tried to take advantage of frustrations that immigrants often experience in the US by stepping up efforts to skim from its neighbour's talent pool. In 2023, it introduced the 'Tech Talent Strategy' which made it possible for anyone with a US work permit to also become eligible for a Canadian work permit as well. The quota of 10,000 visas was exhausted within forty-eight hours.

'The US is not doing itself a favour with its visa policies,' Norbert Lütkenhaus, the Executive Director of the Institute for Quantum Computing in Waterloo, who immigrated from Germany, told me. 'I mean, thank you very much. Each time the US steps on the brake there we get more faculty members, post-docs, and graduate students here – they come to Canada instead of the US.'

This laissez-faire approach extends to talent that might be shut out from the US because of political reasons. Lütkenhaus says that the first wave of these people came from mostly Muslim-majority countries right after 9/11. Now its shifted to places like China, Russia and Iran.

A notable recent case is that of Sara Sabour, a young researcher from Iran who after receiving her undergraduate degree in computer science from Sharif University of Technology in Tehran in 2013 was accepted to graduate school at the University of Washington. But the US denied her visa, deeming that her field of study, computer vision, could have military applications. She eventually found her way to the University of Toronto and then to Geoffrey Hinton's research group. The two have collaborated closely on capsule networks, a newer type of neural network that can give machines the same three-dimensional perspective as humans. Sabour now works as a research scientist at Google in Toronto.

'It's the same thing about conferences,' continues Lütkenhaus. 'Now people are wondering: should I do a conference series in the US if I'm not sure whether my key speaker can actually get a visa to talk because he might come from China or from India? And suddenly we have a big empty slot because the key speaker did not get his visa. So sometimes we wonder exactly how people are thinking in the US.'

Getting more international students to come and study in Canada has also been an important part of the country's strategy to up its immigration numbers. Canada hosts the third largest cohort of international students in the world after the US and UK, over half a million students. One in eight of all students who leave home globally to study in another country leave home to study in Canada. And most of them don't plan to leave. According to a poll by the Canadian Bureau of International Education, over 60 per cent of the respondents said that they want to stay in the country after graduating.

'We recently had the visit of the German Chancellor and the Canadian Prime Minister at Mila and one of the German students was explaining that specific idea to the German Chancellor who asked, why did you leave Germany? Why come to Canada?' recalls Valerie Pisano. 'He said, just look around. What you have here doesn't exist in Germany. Look at how vibrant this is. I'm 25 years old and I can be a part of this. The godfather of artificial intelligence is here. Look at my peers, look at the people I get to work with.'

9

Canada has been quietly building out its tech industry for the past decade and the world is starting to pay attention. In 2023, it had twenty-five private tech companies valued at over a billion dollars and ranked eighth among countries that produce such companies. In 2022, over $10.5 billion was invested across 706 venture deals in Canadian tech startups, up from $1.5 billion a decade ago.

Foreign investment is pouring in: two-thirds of the capital invested in Canadian tech companies comes from outside Canada.

'I just think capital follows companies, not the reverse. Everybody thinks capital is the thing that starts these ecosystems, but that is incorrect,' says Damien Steel, former Managing Partner and Global Head of Ventures at OMERS Ventures. 'The money follows where the opportunities are and so to me that was just a great validation for the Canadian ecosystem, that hey, we're doing something right here.'

Twenty years ago, two companies loomed large over the Canadian tech scene: Research in Motion (RIM), the maker of BlackBerries, and Nortel Networks, the backbone for the telecoms infrastructure in many parts of the world. Neither could stay ahead of more innovative competitors overseas and both met an abrupt and precipitous end.

'They weren't organizations that promoted innovation, they did the opposite,' says Steel. 'RIM was notorious for trying to squash innovation from leaving its organization. Today that's not the case. You look at Shopify and the stuff they do, they actively promote their angel groups within Shopify that are funding startups and entrepreneurs that leave Shopify to start something. That's what Silicon Valley is: success breeding success and supporting itself as an ecosystem. We've just hit that.'

Canada's tech economy is no longer about one or two big names. It has lots of players with different roles working together to sustain a virtuous cycle of successful companies creating more successful companies. Universities like McGill and the University of Toronto make research breakthroughs and train talent. Incubators like CDL get good ideas off the ground. Alums from breakout success stories like Shopify go out and write their own breakout success stories. And institutes like Vector, Amii and Mila bring in government support to tie it all together.

The most significant shift though might be the mindset. 'I just think as an ecosystem,' says Damien Steel. 'We need to have a broad enough view and understand that we win as the ecosystem wins.'

But sceptics would point out that despite all the talk about Canada making a play for AI supremacy, its companies are still relatively new, small and practically unknown outside the country. When people think AI they think OpenAI, Nvidia, DeepMind or ByteDance. Canada has not yet produced an AI company that reaches even a fraction of their scale. Ask people outside Canada to name a Canadian AI company and most would be hard-pressed to name even one.

So as much as Canada gets all the points for trying, and as much as the values that underpin its tech economy might bring forth positive sentiments and warm feelings, the question remains: can a Canadian AI company ever be as consequential as the big leagues overseas?

'I have absolutely no doubt this is going to happen,' says Sanja Fidler, Director of AI at Nvidia.

Sanja, who is also an associate professor at the University of Toronto and was one of the co-founders of the Vector Institute, says it's unfair to compare Canadian tech companies with some of the more established names who have been around for much longer, and it's only a matter of time before a Canadian AI company like Cohere, Waabi or Deep Genomics joins American, British and Chinese titans at the top.

'The people running these companies were heavily recruited by big tech giants previously and they're some of the smartest people I know,' says Fidler. 'The most talented in tech advancements in AI. Just looking at the momentum and trajectory of some of these companies, there's no doubt that they're going to be at the same level.'

CIFAR predicts that by 2030 Canada will have one of the most advanced national AI ecosystems in the world. The Tortoise Global

AI Index, which measures AI investment, innovation and implementation, places Canada eighth in the world.

'Toronto now by some accounts has as many AI startups as any city in the world starting from virtually zero five years ago,' says Jordan Jacobs. 'And we want to make sure that in fifty years when we look back, Canada is one of three hubs. I assume the US will be a hub, and China will be a hub – and Canada will be the third hub.'

CONCLUSION

Social Animals

*'Cultures are not like biological systems;
cultures are biological systems.'*

I

When I first started writing this book, the idea was to explore whether US supremacy in new technologies is waning. The moment seemed to be right for such an enquiry. The question was fodder for much commentary globally and found its most erudite explication in Alan Greenspan and Adrian Wooldridge's 2018 tome *Capitalism in America* in which they argued that the economic dynamism that set the US apart from other nations was now beginning to fade: 'Look at any measure of creative destruction, from geographic mobility to company creation to tolerance of disruption, and you see that it is headed downward. The United States is becoming indistinguishable from other mature slow-growth economies such as Europe and Japan in its handling of creative destruction.'

The parallel rise of China only brought this decline into sharper relief. The country is beginning to take on the aspect of what the US was a generation ago: that one place that eclipses all other places. Today China makes more solar power, has more robots working in its factories and more EVs roaming its streets than the rest of the world combined. It doesn't take much imagination to see the US as the established incumbent struggling to hold things steady and China the fast-growing disrupter for which every graph is up and to the right.

But as I looked closer at what's happening elsewhere it did become hard to avoid the conclusion that predictions of American decline, at least in the realm of new technologies, are vastly exaggerated. There's as much evidence that its innovation engine is accelerating as there is that its spluttering to a halt. The combined market value of the ten largest US tech companies exceeds the entire GDP of every single country in the world except the US itself. And that ascendancy is not a holdover from a bygone era, it is very much a contemporary development. Apple became the first American tech company to cross a trillion-dollar valuation in 2018. Seven others have joined that club since, some crossing that mark multiple times over. The rest of the world, the whole world, doesn't even have one. What more needs to be said?

The only concession that could perhaps be made to the rest of the world is that twenty-five years ago the technology landscape was dominated almost entirely by the US; now the US is still in a commanding lead and then there are a handful of other countries, twenty, tops, that are also gaining relevance. The US is growing, China is growing faster, others are growing too but not as fast as the US and China.

And apart from China it doesn't seem like the rest of the world has a particularly competitive relationship with the US in tech. Many position themselves as extensions of Silicon Valley rather than its antagonists. In South Korea and Sweden, it's aspirational for startups to have a dual presence: one foot in the US and one in their home market. They want to be seen in equal parts as American and wherever they are originally from.

Even Chinese executives would off the record tell me about how truth be told exposure to the US was still very important to their businesses. Maybe not as a market in which to do business in, that was too risky and they could survive just fine without, but as a place to socialize themselves in the state of the art in tech and pick up early signals about where things are headed in frontier disciplines like AI so that they could orient their efforts in that general direction as well.

China is the only competitor. Others are trying but they're not good enough to be dangerous. The most successful tech companies outside China whose business model does not rely solely on monetizing their home markets still consider it aspirational to eventually find their way to the US. That may be in the form of an outright acquisition or a public listing or a physical move. But being stateside is still the marker that you've arrived.

Hugging Face didn't stay in France for long and UiPath too outgrew Romania fairly early in the game. Even long-established companies, like Arm in the UK, ditched a domestic listing to go public in the US. As did Spotify and Grab. Most European VCs that I spoke to said they actively pushed their companies to find a big American patron; some encouraged founders to win the US first and then try and find their way back, expanding into Europe instead of expanding out of Europe.

China doesn't seem to have a similar sway outside of its own borders. Chinese companies want market access abroad and other foreign companies want market access in China. But it doesn't really have the same kind of symbiotic relationship with other major tech capitals like the US tech community has in places like Taiwan and India.

Some Chinese companies are actively distancing themselves from their home market. Shein, a fashion retailer which at one point was valued at over $100 billion, de-registered its original company in Nanjing and moved its headquarters to Singapore. Temu moved its headquarters to Boston and its parent company PDD Holdings moved its base to Ireland.

American dynamism may yet fade, but it would have less to do with factors intrinsic to its tech industry and more to do with its fractured polity and backlash against a highly successful tech industry and the inequities it has wrought, a consequence ironically of too much innovation and not too little. China's liabilities are also in plain sight. The most notable being that its tech industry seems to have outgrown the political frame within which it operates. The most vivid demonstration of these frictions came in 2021 and it remains to be seen whether the country's entrepreneurial class has been subdued for good or if it will rattle the cage from the inside again.

The rest of the world is still figuring out how to innovate. US and Chinese companies seem to have solved that puzzle. Their dynamism is constrained less by an inability to innovate and more by resentment at the vast new powers that cracking that problem has conferred upon them.

An interesting statistic to ponder is that in 1990, China was responsible for only about 1.8 per cent of global GDP. It has since grown tenfold and stands at 19 per cent today. Compare that to the US which was responsible for about a quarter of the world's GDP in 1990. In 2024, it still produces a quarter of the global GDP. China has expanded but so far not at the expense of the US. It's the other rich world economies that have lost out in relative terms, to both the US and China. In 1990, the US accounted for 39 per cent of the GDP of G7 nations. Its share now stands at 57 per cent. I would hypothesize the relative development of tech sectors of the US, China and the rest of the world would mirror that broader economic trajectory.

2

A comparative study of what's going on in different places around the world not just in technology but in all sorts of contests – sports, economics and the rest – is useful on two levels. The first is the obvious one and that's just the national competition element of it which makes good headlines and gets those old tribal instincts revved up; that is to say who's doing well and who isn't and what does the scoreboard look like.

But the more utilitarian approach would be to look at these tournaments less through the lens of national competition and more as a test of the relative merits of competing systems. All societies are working through similar challenges – growth, governance, sustainability – and the world is a big enough place to experiment with different approaches in different places to see what works and what doesn't.

And this is already happening for the big questions. So, if the desired outcome is a good society, then there are all sorts of contrasting approaches being tested out, everything from free market liberal democracies to different shades of socialism all the way through to communism and more theocratic forms of ordering things. Regardless of personal preferences, there is value to having these trials run in parallel, even some of the crazier ones, to see which ones yield better outcomes than others.

In business too there's a diversity of approaches. There's the *keiretsu* system in Japan which has large groupings of interconnected companies which have extensive cross-shareholding typically centred around a major bank. Mitsubishi would be a good example. Then there's the German Mittelstand which we've discussed in these pages, the smallish SMEs that are the backbone of the country's industrial economy.

China relies disproportionately on its massive state-owned enterprises like Sinopec and the China Construction Bank which eclipse the country's biggest private companies by orders of magnitude. And then there are the hybrid *chaebols* in Korea where it can be hard to tell where the state ends and private initiative begins. Different places do business very differently.

I was expecting a similar story in how various countries systemically build out new technologies; Sweden does it this way, China does it that way, India does it the other way. And there could be lessons for others in these approaches. That largely turned out not to be true.

While the world's older industrial economies might have looked very different, its newer tech economies are strikingly similar and becoming even more similar over time. Every country's strategy relies on some variation of having a DARPA-like agency on the state side and then venture-backed startups in the private sector.

These similarities run all the way down from the structural elements to the feel of the cultures of these companies and more stylistic aspects like those brightly coloured t-shirts they wear and even the linguistic quirks that mark how this subclass of the species interacts with each other. Delphic utterances like bruv, bruvsky, brotato, broham, brochacho, brofessor and Broseph Stalin which would mean little to the uninitiated are the shared symbology and tribal markers that is the veritable social glue that binds together this herd of contrarian thinkers from New York to Stockholm to Tokyo.

IKEA and General Electric were very different; Spotify and Stripe are more similar. That old diversity of doing things has given way to more of a monoculture of thought and action. There is ironically very little experimentation in how the newer generation of companies is run in different places: whether it's Klarna in Sweden or Mistral in France or Grab in Singapore, things are a bit same same.

The contest is not between different models but different intensities of the same model. China is different politically but on the economic front, especially in the tech industry, its more American than the Americans. The buzz of places like Shenzhen can feel like Silicon Valley on steroids. The gladiatorial competition among new ventures which Kaifu Lee valorizes in his book can only be described as a form of hyper capitalism, a more intense and concentrated variant of the ideology which would be regarded as too extreme even in its Western cradle. The difference is not of kind, it is of degree.

I asked David Allemann, the founder of ON, probably the hottest new venture to come from Switzerland, whether his company is different because of where it's from. ON is unusual in that it has two CEOs who share power at the top and I wondered if that was at all inspired by the communitarian ethos that animates the country more generally. The Swiss political system places a high premium on dispersing power both horizontally and vertically; the country doesn't even have a head of state and the highest executive authority rests with a council of seven people. Allemann thought the similarity was coincidental. 'I don't think we're that different just because of the fact that we went global so early,' he said.

So there is more convergence to the American way of doing things instead of something that is meaningfully different in different places, an underwhelming end of history as applied to building out new tech. There is talk of other approaches, like DeepMind exploring the idea of becoming a public interest company or Jack Dorsey suggesting that Twitter should have been a protocol, something more like email and less like a commercial social network. OpenAI too toyed with the idea of being a nonprofit before it lapsed back into an all too typical Valley company.

And then there is a growing movement that more technology should be developed as a global public good, similar to how the internet was created as an open and freely accessible platform and not as a commercial product. But there's little evidence that anyone's working on something that is systemically different at that level, a radically alternative approach to doing tech that could be the foil to whatever is happening now.

There's two ways to compete. The first is to do whatever the others are doing and try and do it bigger, better, faster. The other is to do it differently. Most of what's going on right now falls into the former category.

In one way the sameness is not such a bad thing, it just means that in a world otherwise fraught with divisions along just about every conceivable fault line there is something close to a globally accepted standard in one important area of activity. But in other ways it's dispiriting. One would assume that with all the talk about thinking from first principles that one hears so often in technology circles there would be a similar enthusiasm to reimagine the whole enterprise at the most basic level that would fundamentally break the mould for how new technologies are brought to life and how they are put to use.

3

If different places are becoming more similar from a technology creation perspective then does location not matter any more? In recent years, there's been a lot of talk about how great companies can come from anywhere and it's not necessary for startups to cluster in one place as much as they needed to before. The knowledge, infrastructure and talent that go into building companies are now much more dispersed.

There have also been hypotheses that Silicon Valley has moved to the cloud and the internet has democratized entrepreneurship and made it so that everything doesn't need to happen all in one place. The rise of remote working and tech workers' preference for a more mobile lifestyle and less crowded metropolises have, some say, all moved things in the direction of hubs just not mattering as much as they used to.

'Tech hubs are losing the talent war to everywhere else,' wrote the *Wall Street Journal* in 2023. The author noted: 'As the tech industry as a whole has matured, it is following the pattern of every previous industrial revolution: knowledge, capital and talent is spreading out, across America and the world.'

I think the jury is still out on that one. Based on my discussions it seemed that location still matters. Knowledge and resources might be available in more places, but hubs were never about just that. Their principal function is to create a hive of like-minded people who are all working on similar problems and can benefit from serendipitous encounters and supercharge their learning based on all the experience and competitive spirits floating in that environment. 'If you ski with the best you get better fast; if you need to learn everything by yourself it's much more effort,' Nathalie Casas, a senior executive with the Swiss Federal Laboratories for Materials Science and Technology, told me.

Entrepreneurship is still not as normalized as we would like to think it is, and a big proportion of the technically trained workforce even in developed Western economies would still prefer to work in stable jobs in big companies. That's the story of Germany where the top engineering grads would still rather work for BASF or BMW than take the risk of starting their own ventures.

Switzerland has a high concentration of top-tier talent in applied physics and related engineering disciplines because of the presence of these cathedrals to big science like CERN, but there's little evidence that enough of them leave to start their own companies or go to work for other startups. That technical talent would much rather stick to the stability and security of their day jobs. The talent is there, the money is there, the infrastructure is there, but entrepreneurship isn't in the air so to speak.

Hubs make it socially acceptable to be a founder and entrepreneurs need not become social outcasts during the rough patches that everyone inevitably has to endure. John Kim, the founder of

Sendbird, a software developer that makes APIs for platforms like Reddit and Paytm, told me how he went from feeling like an oddball in Korea to feeling like he belonged the moment he landed in the Valley: 'Every café I went to, there was a startup pitching to a VC, everyone was talking about technology literally all around me. And it was so shocking and also so comforting in a way. I felt like I found my home, this is where I wanted to be.' It's impossible to replicate that feeling on a Zoom call.

And that's largely why even the headline examples of startups that came from 'nowhere' eventually relocated to more mainstream locations. Up to a fifth of the startups that have raised more than €1 million – a minuscule amount of money in the larger scheme of things – in the Central and Eastern European region eventually migrate abroad; this includes superstars like UiPath, Grammarly and Productboard, all $1 billion-plus companies. According to one study, over $50 billion in enterprise value has flowed out of the region. The most ambitious companies with the brightest futures just don't want to miss out on the network effects being spun in more established tech hubs. Colocation matters.

Which leads to the final question that if location still matters and there are only a handful of places which provide a suitable environment for tech entrepreneurship then are there any general conclusions that can be drawn about what makes these places so different?

The notion that most people have is that permissive settings are more suitable environments for this sort of thing than their more controlled counterparts; that loose cultures drive better outcomes than tight ones. 'Innovation is the child of freedom and the parent of prosperity,' writes Matt Ridley in his book *How Innovation Works: And Why It Flourishes in Freedom*. And that sentiment does at first satisfy some deeply held sense that we have of what it's all about.

But there's obviously a China-sized hole in that argument. It's uncontroversial to say that the country is heavily dirigiste compared to its counterparts in the West. And yet in tech it's outperforming just about every country except the US. Chinese innovation is driven by factors precisely the opposite of freedom and prosperity; Kaifu Lee talks about the maniacal work ethic that powers their tech industry which is driven in large part by the grinding poverty of the average Chinese worker that often stretches back generations.

If freedom and prosperity are what made innovation work then there would be a much better correlation among how free and rich countries are and how innovative they are. But the question is interesting and worthy of closer examination precisely because that relationship simply does not hold under closer examination.

If it were such a simple matter, we would have seen the intensity of tech innovation move from North America to Western Europe, through the Pacific and Latin America and only then to Asia. But it's not like that. Sweden, Denmark, Norway, Finland and Iceland are very similar in their socioeconomic attributes, but Sweden is the clear outlier when it comes to driving innovation outcomes.

So the question of what makes different places fertile grounds for new things is a lot more complex than it just being a function of liberty and wealth. Historically too, some very tightly controlled regimes have nevertheless shown tremendous technological competence: the Soviet Union before the fall of the Berlin Wall and Germany and Japan before World War II would be the most obvious examples.

Having freedom and prosperity is better than not having them. The argument is not that they don't matter. The argument is that the relationship between them and the capacity to make new things is not as straightforward as maybe our instincts would lead us to believe. They are very often the same reasons that are given for why some places lose their drive and go soft and complacent.

In some rich environments, the economic waters are too placid for strong currents of technological change to work their magic; the animal spirits just aren't there. Japan has been consistently free and prosperous for the past five decades. It's even freer than the US according to rankings issued by Freedom House, scoring a near perfect 96 on its benchmark compared to the more modest 83 for the US. But Japan's relevance to tech innovation has gone off a cliff in the recent past.

The question of why some places are more innovative than others defies easy explanation and is not tractable to grand theorizing. Which is precisely why going deep into individual examples is a worthwhile undertaking which is what we've attempted to do here.

A closer look at these examples brings two broad categories of explanation to the fore. The first systemic and the second cultural. The systemic explanation accounts for differences in institutions, rules and procedures: structural elements like how higher education is organized, how research is incentivized, and how companies are formed and funded.

But the systemic explanation is incomplete because you can have similar systems in many different places and still not get similar outcomes. This is where culture comes in. We've known at least since Weber wrote about the Protestant Ethic that different behavioural attributes of different social groups can drive different economic outcomes.

But it would be useful to look at human cultures not just in the sociological sense but in the more biological understanding of the term. Not unlike the stuff that scientists look at in petri-dishes.

We tend to think of human beings as the most complex organisms on the planet, but that sense of priority might be misplaced. A collection of individuals forms a culture, and that culture is a living, breathing entity that displays collective intelligence and intention. Seen from this lens, different cultural groups are not just national identities but distinct social organisms.

Few English persons who were alive a hundred years ago are still alive today. And yet the attributes that we consider English have remained remarkably consistent over time. Every single member of a culture can perish, but the culture survives; not unlike how the human body replaces virtually every cell every few years but the larger organism retains consistency of self.

Biologists write about superorganisms like ant colonies, microbiomes and swarms composed of relatively simple organisms like ants, bacteria and bees which come together and organize into something much greater than the sum of their parts. The larger entity constitutes a separate superorganism the workings of which might not be in the awareness of its constituent parts.

There's no reason to believe that something similar is not happening with human cultures; after all societies are not *like* biological systems, societies *are* biological systems. Seen from this lens the surface of the planet is essentially smeared with different blob-like cultures which act with collective intelligence and intention: an individual merely a fragment of its larger culture. Most of what we call international relations is just these larger social animals hissing and biting at each other. Societies writ large form a kind of social ocean, with an individual a sort of droplet in that ocean. Some of these oceans are turbulent with vigorous flows while others stay placid and subdued.

It would be appropriate to look at the variations in the capacities of different cultures to bring newness into the world as a kind of technological fertility. Some cultures are particularly fecund, constantly giving birth to new ideas and innovations, while others lie dormant; some fizz and pop while others stay inert and quiescent. Understanding innovation, then, requires us to look beyond mere systems and structures to the internal properties of the cultures from which technologies spring forth.

I drive a Volvo XC40, an SUV which in its no-nonsense practicality is sometimes seen as more than just a car and something of a receptacle of Swedishness on wheels; a modern-day version of the mosaic tiles of the Roman Empire or ceramic vessels of Ancient

Greece, everyday objects that carry the imprint of an entire culture as if to say: this is who we are.

The crossover is keenly self-aware of its totemic status; woven into the seam of the front passenger seat, sticking out like a garment tag, is a tiny blue and yellow Swedish flag. Volvo doesn't label its cars 'Made in Sweden', it labels them 'Made by Sweden'. It's an expression of national pride. But also a deft confession that the company doesn't make all its cars at home any more.

But in a deeper sense it was always true that all technologies are not products in their environment but products of their environment, oozing out of and continuous with the bubbling social ferments in which they germinate and bloom.

ACKNOWLEDGEMENTS

Writing a book is a hard and solitary endeavor but it would've been immeasurably harder and more isolating without the support and participation of many others who made this work possible. I would like to thank the *Financial Times* and McKinsey & Company for launching the Bracken Bower Prize to get new voices into business literature. As a first-time author I really could not have hoped for a smoother entry point into getting my words into print; I dipped a toe in the water and before I knew it I was up to my neck.

My gratitude to the entire BBP community. Andrew Hill for making room for this initiative at an institution with interests as broad and varied as the *FT*. Alexandre Lazarow and Michael Motala for giving me the benefit of competing with the best. Colette van der Ven for being one of the few people I could talk to about ideas and writing. And Chris Clearfield for introducing me to the best agent in the world. James Pullen at the Wylie Agency was the first person who made me feel like I had what it takes to go pro, and his sustained enthusiasm carried this book through its many iterations.

I will never forget going to auction in the UK and on the last day going to bed thinking we were moving forward with one publisher only to wake up the next morning to find out that another had swooped in with a last-minute winning bid. I have often wondered how different this book would've been without that entirely serendipitous eleventh-hour switch. I would not rather have any other imprimatur on the jacket than William Collins and they proved to be just the ideal home for this project. Arabella Pike gave me every freedom to explore doing this my way and then stepped in just when the time was right to push me to get this over the line. Her thoughtful comments on the manuscript which came on a Sunday afternoon made me feel like there was at least one other person who cared about this book as much as I did. I am grateful to the entire team at William Collins, especially Freya Alsop and Laura Meyer, for guiding the book through production. Ben Loehnen at Simon & Schuster was generous in picking up the US rights with the words 'I suspect he'll write a dazzling book, and that he'll be writing for years.' That prophecy would prove to be at least partly true in that I would be writing this book for years and though he might not have meant it that way and would indeed have strongly resisted such interpretative latitude I remain grateful to him for his patience and for seeing the project through to publication.

I would like to thank my mother, to whom this book is dedicated, for being the reason why I do anything. My father for being there when it matters. My brother and sister I am generally unsentimental about and their contribution to my life, and this world, lies principally in them making up for their own shortcomings, which are substantial and wide-ranging, by offering up nieces and a nephew who exhaust my capacity for superlatives, so I am compelled to mention them here. My brother-in-law and sister-in-law were in the mix too. For the longest time my niece Fafi would start every conversation with the question: 'How many pages have you written?' There were long stretches of writing and rewriting this book when that number only ever went backward. If she has filled my life with boundless joy I have filled hers with realistic expectations of what progress looks like in this world and for that precious gift, I am sure, one day she will be thankful, even if it might be hard for her to appreciate its value now and much as she would've rather preferred to have been compensated in *Frozen* memorabilia instead. Hoor, Zany, Hannah, and Muhammad are too young to satirize or attempt to profit from my literary angst but I expect that in due course they too will behold what passes for my writing process with childlike wonder and general confuzzlement. My thanks also to Emad Nadim for reasons that cannot be disclosed in polite company.

This book would not be possible without the hundreds of people who agreed to be interviewed; many have appeared in these pages, others have gone unnamed, some of whom would have preferred it that way.

And finally, Kirsten Alisha Laura Williams, empress of Herbolzheim and Kirchzarten, keeper of the faith, for being the signal in the noise, the only person who knows how hard this was to pull off from the beginning to (almost) the end. You like because, you love despite.

NOTES

Introduction: Capitalism with an Emoji Face

1 *These guys hadn't raised any money*: Ilkka Paananen, interview with the author, 14 June 2023.
2 *the highest grossing game in the US*: David Curry, 'Clash of Clans Revenue and Usage Statistics', 22 January 2025, https://www.businessofapps.com/data/clash-of-clans-statistics/
2 *brought in over $10 billion in lifetime revenue*: Jeffrey F. Rayport, George Gonzalez, 'Supercell 2.0: Clash of Plans', Harvard Business School Case Study, March 2024.
2 *more money than the top three highest grossing movies ever*: *Variety*, 'The 30 Highest-Grossing Movies of All Time', 8 April 2024.
3 *all the Harry Potter books, combined*: Statista, 'You're a wizard at making money, Harry', 15 Nov 2018, https://www.statista.com/chart/16114/harry-potter-franchise/
3 *top hundred music singles streamed ever*: ChartMasters, 'Most Streamed Tracks of Spotify', 13 February 2025, https://chartmasters.org/most-streamed-tracks-on-spotify/
3 *valued the company at over $10 billion*: Richard Milne, '"Clash of Clans" maker to "take more risks" in search of billion-dollar hit', *Financial Times*, 14 February 2024.
3 *eleven-digit valuation*: William R. Kerr, Benjamin F. Jones, Alexis Brownell, 'Supercell', Harvard Business School Case Study, October 2016.
3 *€400,000 loan*: Dean Takahashi, 'How the Finns built their gaming startup hub in Helsinki', VentureBeat, 13 November 2013.

4 *It became a geography question*: Ajay Agarwal, interview with the author, 3 October 2022.
5 *The space stream at CDL*: Chris Hadfield, interview with the author, 15 November 2022.
6 *I thought Xavier was pulling my leg*: Roxanne Varza, interview with the author, 18 September 2023.
6 *I like to compare a researcher in Harvard*: Romain Dillet, 'Emmanuel Macron meets with the French tech community', TechCrunch, 9 October 2018.
7 *listed with an emoji*: Kenrich Cai, 'The $2 Billion Emoji: Hugging Face Wants To Be Launchpad For A Machine Learning Revolution', *Forbes*, 9 May 2022.
9 *Top ten countries by number of private billion-dollar tech companies*: Dealroom, 'Unicorns', Retrieved 13 February 2025, https://dealroom.co/guides/unicorns
10 *Top ten countries by volume of venture dollars invested*: Dealroom, 'The State of Global VC', Retrieved 12 February 2025, https://dealroom.co/guides/global
11 *Total tech market cap ($T) per region, 2021–2023*: Atomico, 'State of European Tech 2023', pg 240.
12 *Highest ranking countries in the WIPO Global Innovation Index*: World Intellectual Property Organization, 'Global Innovation Index 2023', pg 19.
13 *The EU is losing the race for innovation*: European Policy Analysis Group, 'EU Innovation Policy: How To Escape The Middle Technology Trap', April 2024, pg 3.
13 *too large a proportion*: John Maynard Keynes, 'The General Theory of Employment, Interest, and Money', pg 149.
13 *unpacks all that mathematical rigour*: Stephen M. Walt, 'Rigor or Rigor Mortis? Rational Choice and Security Studies', Security Studies, Spring 1999.

Chapter One: The Precocious Student

17 *sold half of Tencent to Naspers*: Loni Prinsloo, 'Tencent's 60,000% Runup Leads to One of the Biggest VC Payoffs Ever', Bloomberg, 22 March 2018.

17 *in the early days I used to do speeches*: David Wallerstein, interview with the author, 6 March 2024.
18 *Naspers started as a modest newspaper business*: Alexandra Wexel, 'How a Small Bet on Tencent Made an African Firm One of the World's Most Valuable', *Wall Street Journal*, 30 November 2017.
20 *Tencent has stakes in over a hundred gaming studios*: Steven Messner, 'Every Game Company That Tencent Has Invested In', PC Gamer, 21 November 2024
21 *Spotify of China:* Yue Wang, 'Tencent Music Is Better Than Spotify At Making Money, But Growth Uncertainties Still Loom', *Forbes*, 18 December 2018.
21 *highly diversified global holding company*: Quentin Webb and Jin Yang, 'China's Tencent Becomes an Investment Powerhouse, Using Deals to Expand Its Empire', *Wall Street Journal*, 3 March 2021.
22 *$1.78 billion to buy 5 per cent of Tesla*: Tim Higgins and Anne Steele, "Tesla Gets Backing of Chinese Internet Giant Tencent," The Wall Street Journal, March 29, 2017.
22 *Tencent had amassed a listed investment portfolio*: Primrose Riordan and Ryan McMorrow, "Tencent pursues quieter investment strategy amid China's Big Tech crackdown," *Financial Times*, January 19, 2022.
22 *Wallerstein's job was no longer just about doing well*: Alyssa Abkowitz, 'The Man Who Bets Tencent's "Moonshot" Money', *Wall Street Journal*, 8 February 2018.
23 *inbound foreign investment into China*: Cheng Leng, 'Foreign direct investment in China falls to lowest level in decades', *Financial Times*, 19 February 2024.
23 *outbound foreign investment from China*: Yi Wu, 'China's Outbound Investment: Recent Developments, Opportunities, and Challenges', China Briefing, 13 November 2023.
24 *The DoD would eventually drop IDG*: Sarah McBride, 'IDG Capital Dropped from Pentagon's China List', Bloomberg, 17 December 2024.
24 *companies that were extremely capital intensive could look to China*: Shahin Farshchi, interview with the author, 30 June 2023.

24 *We think China is*: Keith Rabois, interview with the author, 23 December, 2022.
26 *today's deglobalized world*: Ingrid Lunden, 'Neura Robotics picks up $55 million to ramp up in cognitive robots', TechCrunch, 19 July, 2023.
26 *so the challenge with China*: Hussein Kanji, interview with the author, 17 August, 2020.
26 *strategic autonomy*: 'European defence, strategic autonomy and NATO', European Parliament Briefing, 23 February, 2024.
27 *It was a lifesaver*: Hermann Hauser, interview with the author, 22 January, 2024.
28 *We sort of neutral countries*: Ilkka Paananen, interview with the author, 14 June, 2023.
31 *China used to produce*: Geoffrey Hinton, interview with the author, 13 October, 2022.
32 *The old science world order*: *The Economist*, 'China has become a scientific superpower', 12 June, 2024.
33 *the single most important*: Matt Sheehan, 'Who Benefits from American AI Research in China?', Marco Polo, 21 October, 2019.
34 *He passed the exam*: Kristi Heim, 'Ya-Qin Zhang, Microsoft's leader in China, prospers in a changed nation', *Seattle Times*, 8 August, 2008.
35 *Money wasn't really an object for me*, Ya-Qin Zhang, interview with the author, 21 August, 2023.
36 *Born out of national humiliation*: *Economist*, 'Tsinghua University may soon top the world league in science research', 17 November, 2018.
36 *According to the Leiden Ranking*: Caroline Wagner, 'China's universities just grabbed 6 of the top 10 spots in one worldwide science ranking – without changing a thing', The Conversation, 3 April, 2024.
36 *In the* Nature *Index*: Bec Crew, 'Nature Index 2024 Research Leaders: Chinese institutions dominate the top spots', Nature Index, 18 June, 2024.
36 *Tsinghua is overwhelmingly stronger*: Simon Marginson, interview with the author, 2 July, 2024.
37 *According to a paper*: Yongjun Zhu, Donghun Kim, Erjia Yan,

Meen Chul Kim, and Guanqiu Qi, 'Analyzing China's research collaboration with the United States in high-impact and high-technology research', Quantitative Science Studies, 8 April, 2021.

37 *Microsoft has been under pressure*: Karen Weise, Cade Metz and David McCabe, 'Microsoft debates what to do with AI lab in China', *New York Times*, 11 January, 2024.

38 *Even in the United States*: Jie Tang, interview with the author, 21 January, 2024.

38 *It is backed by*: Eleanor Olcott, 'Saudi fund invests in China effort to create rival to OpenAI', *Financial Times*, 31 May, 2024.

39 *I think it's very difficult*: Rui Ma, interview with the author, 5 April, 2023.

39 *According to the US State Department*: Lily Kuo and Kate Cadell, 'Chinese students, academics say they're facing extra scrutiny entering US', *Washington Post*, 14 March, 2024.

42 *So the transformation China went through*: Kaifu Lee, interview with the author, 21 December, 2023.

44 *Wrapped up in this conviction*: Kaifu Lee, 'AI Superpowers: China, Silicon Valley, and the New World Order', Houghton Mifflin Harcourt, 2018.

45 *Eight out of ten startups*: Fleximize, 'The speed of a unicorn', https://fleximize.com/unicorns/

46 *attacked by a mob*: Wes Davis, 'A crowd destroyed a driverless Waymo car in San Francisco', The Verge, 11 February, 2024.

48 *To have lived in China since 1990*: Zak Dychtwald, 'China's New Innovation Advantage', *Harvard Business Review*, May–June 2021.

50 *Beijing poured $125 billion*: Edward White, 'How China cornered the market for clean tech', *Financial Times*, 9 August, 2023.

50 *We tend to forget*: Dan Harsha, 'Taking China's Pulse', *Harvard Gazette*, 9 July, 2020.

50 *Many Chinese believe*: Rana Mitter and Elsbeth Johnson, 'What the West Gets Wrong About China', *Harvard Business Review*, May–June 2021.

53 *natural, long overdue course-correction*: Lizzie C. Lee, 'Why Did China Crack Down on Its Ed-Tech Industry?', The Diplomat, 5 August, 2021.

53 *China's inequality levels*: Thomas Piketty, Li Yang, and Gabriel

Zucman, 'Capital Accumulation, Private Property, and Rising Inequality in China, 1978–2015', *American Economic Review*, July 2019.

53 *President Xi wrote an essay*: Dexter Tiff Roberts, 'What is "Common Prosperity" and how will it change China's relationship with the world?', Atlantic Council Issue Brief, December 2021.

54 *Some companies ceded*: Ryan McMorrow, Qianer Liu, and Cheng Leng, 'China moves to take "golden shares" in Alibaba and Tencent units', *Financial Times*, 13 January, 2023.

54 *third most advanced in the world*: Stanford Centre for Research on Foundation Models, 'A holistic framework for evaluating foundation models', retrieved 16 February, 2025, https://crfm.stanford.edu/helm/lite/latest/

Chapter Two: Steeples of Excellence

57 *Silicon Valley is over*: Kevin Roose, 'Silicon Valley is Over, Says Silicon Valley', *New York Times*, 4 March, 2018.

57 *Peak Valley*: Alexandra Suich Bass, 'Peak Valley: Why startups are going elsewhere', *The Economist*, 1 September, 2018.

57 *a particularly sharp critique*: Brian Merchant, 'The End of the Silicon Valley Myth', *The Atlantic*, 29 December, 2022.

57 *took to the pages of the Financial Times*: Michael Moritz, 'Silicon Valley would be wise to follow China's lead', *Financial Times*, 17 January, 2018.

58 *Social diversity and common good*: Zelda Bronstein, 'How Silicon Valley Millionaires Stole Progressivism', *The Nation*, 15 July, 2014.

58 *I think the Valley's fundamentally changed*: Victor Hwang, interview with the author, 19 May, 2023.

59 *The book is a meditation*: Victor W. Hwang, 'The Rainforest: The Secret to Building the Next Silicon Valley', CreateSpace Independent Publishing Platform, 2012.

60 *Detroit correction*: Keith Rabois, interview with the author, 23 December, 2022.

61 *Mt Gox was a crypto exchange*: Robert McMillan, 'The Inside Story of Mt. Gox, Bitcoin's $460 million disaster', *Wired*, 3 March, 2014.

Notes

62 *A prominent New York newspaper*: Kevin Roose, "The Doomsday Cult of Bitcoin', *Intelligencer*, 4 March, 2014.
63 *No one knew what happened*: Michael Gronager, interview with the author, 7 December, 2023.
63 *They are now known as*: Gareth Jenkinson, 'Blockchain detectives: Mt. Gox collapse saw the birth of Chainalysis', Cointelegraph Magazine, 27 September, 2023.
65 *Quite frankly*: John Hennessy, interview with the author, 13 February, 2024.
66 *capital of the AI revolution*: Nitasha Tiku, 'AI is reviving San Francisco's tech scene. Welcome to "Cerebral Valley"', *Washington Post*, 15 March, 2023.
67 *I think in AI in terms of fundamental technology*: Ya Qin Zhang, interview with the author, 21 August, 2023.
68 *Four out of five GPUs*: Air Street Capital, 'State of AI Report', retrieved 16 February, 2025, stateof.ai
68 *There's a war going on*: Stephen Witt, 'How Jensen Huang's Nvidia is Powering the AI Revolution', *New Yorker*, 27 November, 2023.
69 *Oh, I think it's going up*: Edouard Bugnion, interview with the author, 13 February, 2024.
70 *Over half of all billion*: Stuart Anderson, 'Most Billion-Dollar Startups in the US Founded by Immigrants', *Forbes*, 26 July, 2022.
70 *Over half of the residents*: Shawna Chen, 'The Californians who speak a non-English language at home', Axios, 7 September, 2023.
71 *You can go to live in France*: Ronald Reagan, 'Remarks at the Presentation Ceremony for the Presidential Medal of Freedom', 19 January, 1989.
72 *In the early stages of a startup*: Farhad Manjoo, 'Why Silicon Valley Wouldn't Work Without Immigrants', *New York Times*, February 8, 2017.
73 *I mean you look at*: Nathan Benaich, interview with the author, 30 January, 2024.
74 *Look, the artists*, Bilal Zuberi, interview with the author, 14 July, 2023.

75 *magisterial twelve-volume history*: Arnold Tonybee, '*A Study of History*', Oxford University Press, 1962.
77 *hard to grasp*: Levi Pulkkinen, 'If Silicon Valley were a country, it would be among the richest on Earth', *The Guardian*, 30 April, 2019.
78 *In the winter of 1883*: Theresa Johnston, 'About a Boy', *Stanford Magazine*, July/August 2003.
78 *He was smiling*: Bertha Berner, '*Incidents in the life of Mrs. Leland Stanford*', Edwards Brothers Incorporated, 1934.
79 *For three weeks*: Herbert Charles Nash, '*In Memoriam: Leland Stanford Jr*', Columbia University, 1884.
79 *The children of California*: 'Stanford History', retrieved 16 February, 2025, https://founders.stanford.edu/stanford-history.
80 *original Palo Alto tech mogul*: Malcolm Harris, 'Palo Alto's First Tech Giant Was a Horse Farm', *The Atlantic*, 8 February, 2023.
80 *Perhaps the greatest sum*: Andrew Carnegie, '*The Gospel of Wealth*', New York: Carnegie Corporation of New York, 2017 (first published in 1889).
81 *Some day you will see Palo Alto*: Hubert Howe Bancroft, '*A History of the Life of Leland Stanford: A Character Study*', Biobooks, 1952, pg 116.
81 *A university, like a tree*: Ibid, pg 117.
83 *What Edison's Menlo Park*: Ibid, pg 291.
84 *cornerstone of the Allied military strategy*: Steven Blank, 'Secret History of Silicon Valley', retrieved 17 February 2025, www.steveblank.com/secret-history
86 *The years after the war*: Ashlee Vance, 'Back from the dead, Silicon Valley icons hitchhike across the US', *The Register*, 4 August, 2006.
86 *No nation can maintain*: Harry S. Truman, 'Special Message to the Congress Presenting a 21-Point Program for the Reconversion Period', 6 September, 1945.
87 *the research scientists of the country*: Vannevar Bush, '*Science, The Endless Frontier*', Princeton University Press, 2021.
87 *military–industrial–academic complex*: Pankaj Mehta, 'Defense Spending: The Endless Frontier', *Jacobin*, 27 August, 2019.

87 *getting defence dollars into the system*: Margaret O'Mara, '*The Code: Silicon Valley and the Remaking of America*', Penguin, 2019.
89 *Every time we came up with an idea*: Rhett Morris, 'The First Trillion Dollar Startup', TechCrunch, 26 July 2014.
89 *It isn't a story simply*: Margaret O'Mara, interview with the author, 1 December 2022.
91 *I argued that Silicon Valley*: Anna Lee Saxenian, interview with the author, 22 November 2022.
93 *The United States companies claim*: Andrew Pollack, 'Japan's Big Lead in Memory Chips', *New York Times*, 28 February 1982.
94 *I was interested in the different theories*: Sebastian Mallaby, interview with the author, 12 June 2023.
96 *Google was a gamble*: Sebastian Mallaby, *The Power Law: Venture Capital and the Making of the New Future*', Penguin, 2022.

Chapter Three: Busting Monasteries

100 *I find it hard to believe*: Ian Hogarth, 'AI Nationalism', 13 June 2018, www.ianhogarth.com/blog/2018/6/13/ai-nationalism.
100 *probably failed*: CNBC, 'DeepMind would have "probably failed" without Google, says early investor', 11 November 2020.
101 *it was really like watching*: Sonali De Rycker, interview with the author, 25 August 2020.
101 *around $22 billion*: HSBC and Dealroom, 'UK Innovation: 2024 forward look', 10 January 2024.
102 *country's relative decline*: Derek Thompson, 'How the UK Became One of the Poorest Countries in Western Europe', *The Atlantic*, 25 October 2022.
103 *caravan societies and citadel societies*: Alan Greenspan and Adrian Wooldridge, *Capitalism in America*, Allen Lane, 2018, pg 389.
104 *in America the percentage of a company*: Hussein Kanji, interview with the author, 17 August 2020.
105 *an indigestible blockage*: Deyan Sudjic, 'King's Cross reborn', *Architecture Today*, retrieved 17 February 2025.
105 *King's Cross was a total wasteland*: Theo Blackwell, interview with the author, 15 November 2022.

106 *if you rubbed a wall*: Andrew Whitehead, *Curious King's Cross*, Five Leaves Publications, 2018.
107 *We define the core area*: Saul Klein, interview with the author, 30 November 2022.
108 *350 tech companies*: Atomico, 'State of European Tech 2023', retrieved 17 February 2025.
109 *In some sense*: Marc Warner, interview with the author, 16 December 2022.
110 *A lot of their technology minds*: Alex Klein, interview with the author, 11 August 2020.
112 *80 per cent of all investments*: UK House of Commons Treasury Committee, 'Venture Capital', Nineteenth Report of Session 2022–23, 18 July 2023.
112 *And that's the negative*: Eben Upton, interview with the author, 7 August 2020.
112 *The UK has some of the highest*: Dan Turner, Nyasha Weinberg, Esme Elsden, and Ed Balls, 'Why hasn't UK regional policy worked?' M-RCGB Associate Working Paper No. 216, October 2023.
113 *economic monopolarity*: John Burn-Murdoch, 'Is Britain really as poor as Mississippi?', *Financial Times*, 11 August 2023.
114 *The contribution in one area*: Royal Commission on the Distribution of the Industrial Population 1937–1940, The National Archives, retrieved 17 February 2025.
114 *derelict urban sites*: UK Government Department for Levelling Up, Housing, and Communities, 'Levelling Up the United Kingdom', 2 February 2022.
115 *My true skill*: Dan Hughes, interview with the author, 6 July 2022.
116 *bottlenecks and slow transaction*: Radix Blog, 'Why blockchains don't scale', 8 February 2018.
117 *Another big breakthrough*: John Colapinto, 'Material Question', *New Yorker*, 15 December 2014.
119 *It's an interesting philosophical question*, Francesco Sciortino, interview with the author, 11 November 2023.
119 *It's too important a technology*: David Barber, interview with the author, 3 November 2022.
120 *preference for a scientific approach*: John Thornhill, 'Huge AI

funding leads to hype and "grifting", warns DeepMind's Demis Hassabis', *Financial Times*, 31 March 2024.
120 *If you want wildly different outcomes*: Matthew Clifford, interview with the author, 27 June 2023.
121 *What's my yardstick of success*: Michael O'Dwyer and Harriet Agnew, 'Jeremy Hunt bets on creating a $1tn "British Microsoft"', *Financial Times*, 13 May 2024.
124 *if national sovereignty in the twentieth century*: John Thornhill 'Arm's destiny vital for Britain's future', *Financial Times*, 30 August 2020.
124 *Now we are about to witness*: Hermann Hauser, 'Arm sale will hit Europe's technological sovereignty', *Financial Times*, 25 August 2020.
125 *I've got to say*: Simon Segars, interview with the author, 31 August 2020.
126 *I'm delighted*: Hermann Hauser, interview with the author, 22 January 2024.
127 *Britain has only three publicly traded companies*: Angus Hanton, 'How America plunders Britain's tech economy', *The Spectator*, 28 April 2024.
129 *The data structures*: Nigel Toon, interview with the author, 2 December 2022.
130 *Too often we have seen*: Tim Bradshaw, 'Graphcore says £900mn UK supercomputer should use its chips', *Financial Times*, 31 March 2023.
131 *The biggest problem is*: Nathan Benaich, interview with the author, 30 January 2024.
133 *Nobody would ask you why*: Jonathan Moules, 'A venture capitalist's European mission', *Financial Times*, 24 April 2010.
133 *What is Venture Capital?*: Madhumita Murgia, 'Europe's startup backer – Neil Reimer, founder, Index Ventures', *Financial Times*, 28 November 2016.
135 *second-guessing*: Hugh Langley and Martin Coulter, 'DeepMind spent years trying to break away from Google. Insiders detail a secret plot sparked by distrust and driven by fears the search giant would sell its AI to the military', Business Insider, 11 September 2021.

136 *agreement signed between the two companies*: Hal Hodson, 'DeepMind and Google: the battle to control artificial intelligence', 1843 Magazine, 1 March 2019.
136 *uncomfortable union*: Parmy Olson, 'Google Unit DeepMind Tried – and Failed – to Win AI Autonomy From Parent', *Wall Street Journal*, 21 May 2021.

Chapter Four: Hyper Gap

138 *less than enthused*: Alex Konrad, 'The Surprise Investors Who Scored Billions From Coupang's IPO', *Forbes*, 15 March 2021.
138 *will deliver the world*: Peter Carlson, '02138: Blushing Crimson', *Washington Post*, 26 September 2006.
138 *breezy lifestyle pieces*: Peter Carlson, 'If you didn't go to Harvard, stop reading', *Los Angeles Times*, 4 October 2006.
138 *Nowhere to Go But Down*: Gabriel Sherman, 'Harvard Prodigy Spends Bradley's $4 Million; Alumni Await Magazine', *Observer*, 19 June 2006.
139 *very short window*: Karen Gilchrist, 'How a Harvard dropout founded South Korea's most valuable startup', CNBC, 3 December 2019.
140 *the most difficult decision*: Karen Gilchrist, 'The most "difficult decision" that helped create a $9 billion startup', CNBC, 5 December 2019.
140 *finally went public*: Choe Sang-Hun and Lauren Hirsch, 'South Korea's Answer to Amazon Debuts on Wall Street', *New York Times*, 11 March 2021.
141 *youngest self-made billionaire*: Grace Chung, 'Coupang's Blockbuster IPO Pushes Founder's Fortune Up To $11 Billion', *Forbes*, 17 March 2021.
142 *stop going overseas*: Son Min-Ho, 'After travel rules relaxed, Koreans took to skies', *Korea JoongAng Daily*, 13 January 2014.
142 *exponential growth curve*: Forest Reinhardt, Sophus A. Reinert, Dawn H. Lau, and Jonathan Schlefer, 'Korea: Miracle on the Han River', Harvard Business School Case Study, 16 February 2023.
142 *global pivotal state*: Yoon Suk-Yeol, 'South Korea Needs to Step Up', *Foreign Affairs*, 8 February 2022.

Notes

142 *shrimp among whales*: Andrew Yeo, 'South Korea as a global pivotal state', Brookings, 18 December 2023.

143 *humanitarian tragedy*: Jong Won Lee, 'The impact of the Korean War on the Korean Economy', *International Journal of Korean Studies*, 2001.

145 *supercharged development*: Benjamin Gomes-Casseres and Seurg-Joo Lee, 'Korea's Technology Strategy', Harvard Business School Case Study, 22 April 1988.

145 *corrupt swine*: Yasheng Huang, 'Korea: On The Back Of A Tiger', Harvard Business School Case Study, 12 May 2002.

145 *family owned businesses*: Eleanor Albert, 'South Korea's Chaebol Challenge', Council on Foreign Relations, 4 May 2018.

146 *even wigs*: The Economist, 'How wigs tell the story of modern South Korea', 27 July 2017.

147 *without them*: Devin DeCiantis and Ivan Lansberg, "A little nut rage is good," The Atlantic, March 13, 2015.

147 *all-encompassing*: Geoffrey Cain, 'Samsung Rising', Crown Currency, 2020.

148 *do-everything monolith*: Chico Harlan, 'In S.Korea, the Republic of Samsung', *Washington Post*, 9 December 2012.

150 *take the top spot*: Ian Sherr and Evan Ramstad, 'Has Apple Lost Its Cool To Samsung', *Wall Street Journal*, 28 January 2013.

152 *Two out of three*: Elsa Lehrer, 'The Chaebol: A Curse in Disguise', *Brown Political Review*, 23 March 2023.

152 *shadowy arrangements*: Carlos Tejada, 'Money, Power, Family: Inside South Korea's Chaebol', *New York Times*, 17 February 2017.

154 *culture of impunity*: The Economist, 'Cases against two ex-presidents of South Korea fit an alarming pattern', 7 April 2018.

155 *net liability*: The Economist, 'The chaebol that ate Korea', 12 November 1998.

156 *hand over household gold*, Frank Holmes, 'How Gold Rode To The Rescue Of South Korea', *Forbes*, 27 September 2016.

156 *causes of the crisis*: Joon-Ho Hahm and Frederic S. Mishkin, 'Causes of the Korean Financial Crisis: Lessons for Policy', NBER Working Paper 7483, January 2000.

157 *a lot of pain*: Yong Kwon, 'The Long Shadow of the Asian Financial Crisis in South Korea', *The Diplomat*, 6 December 2021.

157 *principal villains*: Hansoo Choi, 'Samsung, Lee Jae-yong's Conviction, and How Business in South Korea is Changing', *Harvard Business Review*, 29 September 2017.

157 *In the old economy*: Jimmy Kim, interview with the author, 18 July 2023.

157 *would not have taken place*: Kim Dae-jung, 'Let us open a new era: Overcoming national crisis and taking a new leap forward', inaugural address by the 15th President of the Republic of Korea, 25 February 1998.

158 *platform companies*: Kim Kyung-pil, 'The rise and development of the platform economy in South Korea', *International Journal of Asian Studies*, Cambridge University Press, 4 July 2022.

159 *resonated particularly strongly*: Song Jung-a, 'Kim Beom-su, Kakao: Life of Brian', *Financial Times*, 27 December 2015.

159 *rest of Asia*, Ryan Mac, 'How KakaoTalk's Billionaire Creator Ignited A Global Messaging War', *Forbes*, 27 April 2016.

160 *symbolic passing of the baton*: Ralph Jennings, 'Kakao founder becomes Korea's richest person as shares of his internet giant surge', *Forbes*, 23 June 2021.

161 *175 subsidiaries*: Bohyeong Kim, 'South Korea's Megacorp and super app: Kakao's path to market dominance', Media, Culture, and Society, 16 November 2024.

161 *wrapped into one*: Bryan Pietsch, 'Kakao is Korea's app for almost everything: Its outage forced a reckoning', *Washington Post*, 17 October 2022.

161 *Kakao has turned*: *The Economist*, 'South Korea's government sees tech firms as the new chaebol', 18 September 2021.

161 *tangled in regulatory issues*: *The Economist*, 'Kim Beom-su, the billionaire founder of Kakao, faces trial', 12 September 2024.

161 *arrested*: CNN, 'Founder of South Korea's Kakao arrested for suspected stock manipulation', 23 July 2024.

164 *Our mindset is*: Jae Lee, interview with the author, 21 October 2022.

165 *I created routines*: John Kim, interview with the author, 25 October 2022.

166 *the firm struggled*: Jo Tango and Alys Ferragamo, 'Altos Ventures', Harvard Business School Case Study, August 2022.
166 *I do think*: Han Kim, interview with the author, 15 November 2022.
167 *Building a lab*: Kunwoo Lee, interview with the author, 13 October 2023.
168 *When you are growing up*: Ikkjin Ahn, interview with the author, 25 September 2023.

Chapter Five: Smart Nation

169 *Xiaodong Li*: Shotaro Tani and Kentaro Iwamoto, 'Cash splash: how Sea became south-east Asia's biggest public company', *Financial Times*, 26 March 2021.
169 *not always the smartest person*: Punch Card Investor, 'Sea Ltd, Part 1: Garena – Building a Global Gaming Cash Engine', 31 August 2021, www.punchcardinvestor.substack.com
173 *category of one*: Jessica Tan, 'Game for Garena: Singapore's Answer to Tencent and Alibaba', *Forbes*, 22 July 2015.
173 *Eduardo Saverin*, Alex Konrad, 'Life After Facebook: The Untold Story of Billionaire Eduardo Saverin's Highly Networked Venture Firm', *Forbes*, 19 March 2019.
174 *Sea has hit*: The Economist, 'Sea Group Faces Choppier Waters', 26 February 2022.
176 *finite and limited*: Lee Kuan Yew, 'The Search For Talent', National Archives of Singapore, 12 August 1982.
176 *foreign talent*: Yoojung Lee and Yoolim Lee, 'How Singapore Nurtured Foreign Trio Who Became Billionaires', Bloomberg, 10 August 2020.
176 *transformation in one lifetime*: Charlotte L. Robertson and Mattias Fibiger, 'Singapore: "from third world to first"', Harvard Business School Case Study, 13 May 2024.
178 *59 per cent of the world's tech multinationals*: Singapore Economic Development Board, 'Singapore's Tech Ecosystem', 18 May 2021.
178 *Block71*: Chew Hui Min and Boon-Siong Neo, 'Blk71 – Growth of a Singapore startup ecosystem', Nanyang Technological University, 9 January 2017.

178 *the world's most tightly packed*: The Economist, 'All Together Now', 16 January 2014.

179 *I am often accused*: Alicia P.Q. Wittmeyer, 'Want Your City-State to Become a Capitalist Success Story? Ban Spitting', *Foreign Policy*, 24 March 2015.

179 *leading the charge*: Nitin Pangarkar and Paul Vandenberg, 'Singapore's ecosystem for technology startups and lessons for its neighbors', Asian Development Bank, June 2022.

180 *This skewness underscores*: Wong Poh Kam, 'Global Innovation Hotspots: Singapore's innovation and entrepreneurship ecosystem', World Intellectual Property Organization, 2022.

181 *Many companies that should be closing down*: Toni Eliasz, Jamil Wyne, and Sarah Lenoble, 'The Evolution of State of Singapore's Startup Ecosystem', World Bank Group, March 2021.

182 *better survival rate*: Poh Kam Wong, Ho Yuen Ping, and Ng Su Juan Crystal, 'Growth Dynamics of High-Tech Start-ups in Singapore: A Longitudinal Study', NUS Entrepreneurship Centre Research Report, 2017.

185 *takes institutional form*: 'What is ASEAN?' Council on Foreign Relations, 15 January 2025.

186 *third fastest growing region*: Lightspeed, 'Southeast Asia: Resetting Expectations', September 2024.

187 *China plus one*: Agnieszka Maciejewska and Anton Alifandi, 'ASEAN as a China Plus One destination: Current situation and risk outlook', S&P Global, 25 July 2023.

187 *major player*: Chin Hsueh, 'ASEAN holds the key to reducing US dependence on Taiwan's chip industry', The Diplomat, 1 December 2023.

188 *second home*: Sebastian Strangio, 'Vietnam Signs Agreement with Nvidia to Establish AI Research and Data Centers', The Diplomat, 6 December 2024.

188 *high-profile visits*: Olivia Poh and Suvashree Ghosh, 'Tech Giants Start to Treat Southeast Asia Like Next Big Thing', Bloomberg, 11 May 2024.

189 *maximize relations with both parties*, Kwangyin Liu, Shu-ren Koo, and Silva Shih, 'Five years on, ASEAN a winner in US-China trade war', Commonwealth Magazine, 19 September 2023.

189 *a recent paper*: Gita Gopinath, Pierre-Olivier Gourinchas, Andrea F Presbitero, and Petia Topalova, 'Changing Global Linkages: A New Cold War', International Monetary Fund, 5 April 2024.

190 *whole gamut of activities*: Kaori Iwasaki, 'Chinese Firms Driving Digitalization in the ASEAN Region', RIM Pacific Business and Industries, Vol. XXIII, No. 90, 13 December 2023

191 *warning*: Mercedes Ruehl, 'EU and US warn Malaysia of "national security" risk in Huawei's bid for 5G role', *Financial Times*, 2 May 2023.

191 *wholesale move*: Mercedes Ruehl and Leo Lewis, 'Chinese companies set up in Singapore to hedge against geopolitical risk', *Financial Times*, 30 November 2022.

191 *heightened scrutiny*: James Kynge, Jude Webber, and Christine Murray, 'China's new backdoor into Western markets', *Financial Times*, 5 September 2024.

192 *under pressure*: Mercedes Ruehl, 'Can Singapore hold on to its reputation as Asia's "safe haven"', *Financial Times*, 11 September 2023.

192 *never explicitly*: Abhishek Vishnoi and Yoojung Lee, 'Huawei loses main Singapore 5G networks to Ericsson, Nokia', Bloomberg, 24 June 2020.

192 *more complex*: Chen Nahui and Xue Li, 'Lee Kuan Yew's Legacy for China-Singapore Relations', The Diplomat, 5 December 2016.

193 *benefited from their access*: Shay Wester, 'Balancing Act: Assessing China's Growing Economic Influence in ASEAN', Asia Society Policy Institute, 8 November 2023.

193 *largest trading partner*: Singapore Ministry of Foreign Affairs, 'People's Republic of China', retrieved 19 February 2025, https://www.mfa.gov.sg/SINGAPORES-FOREIGN-POLICY/Countries-and-Regions/Northeast-Asia/Peoples-Republic-of-China.

195 *framework for this strategy*: Siew Kein Sia, Ronald Hee, and Godofredo Ramizo Jr., 'Singapore's Strategic Transformation as a Smart Nation', Nanyang Business School Case Study, 14 November 2023.

195 *coordinated centrally*, Woo Jun Jie, 'Singapore's Smart Nation Initiative – A Policy and Organizational Perspective', Lee Kuan Yew School of Public Policy Case Study, 2018.

195 *in-house*: Orlando Woods, Tim Bunnell, and Lily Kong, 'Insourcing the smart city: assembling an ideo-technical ecosystem of talent, skills, and civic-mindedness in Singapore', Urban Geography, 2023.

199 *salary does make a difference*: Hongyi Li, interview with the author, 8 January 2025.

Chapter Six: Small Wonder

205 *we hope everyone enjoyed*: Shannon Liao, 'NYSE hung the Swiss flag instead of Swedish flag for Spotify's IPO', The Verge, 3 April 2018.

211 *to travel in Switzerland*: Tony Judt, 'In Love With Trains', *The New York Review of Books*, 11 March 2010.

212 *I wish!*: Yoshua Bengio, interview with the author, 5 October 2022.

213 *relinquishing its intellectual property*: Tim Smith, François Flückiger, 'Licensing The Web', CERN, retrieved 19 February 2025, https://home.cern/science/computing/birth-web/licensing-web.

214 *had the technology been*: W3C, 'Frequently asked questions', retrieved 19 February 2025, https://www.w3.org/People/Berners-Lee/FAQ.html.

214 *It was not built*: Christian Rüegg, interview with the author, 9 February 2024.

216 *I was stoked*: Maximilian Boosfeld, interview with the author, 29 December 2023.

217 *If you're a bad student*: Nathalie Casas, interview with the author, 8 February 2024.

218 *It had the Nobel Prizes*: Patrick Aebischer, interview with the author, 23 November 2023.

219 *Mother needs something*: Arnie Cooper, 'An Anxious History of Valium', *Wall Street Journal*, 15 November 2013.

222 *Switzerland is an attractive place*: Edouard Bugnion, interview with the author, 13 February 2024.

222 *I always felt*: Marcel Salathé, interview with the author, 20 December 2023.

223 *DP-3T*: Camela Troncoso, 'Decentralized Privacy-Preserving Proximity Tracing: Simplified Overview', EPFL, 8 April 2020.
224 *This risk*: Andreas Illmer, 'Singapore reveals COVID privacy data available to police', BBC, 6 January 2021.
224 *gives the best privacy*: Christina Farr, 'How a handful of Apple and Google employees came together to help health officials trace coronavirus', CNBC, 28 April 2020.
225 *Maryna Viazovska*: Thomas Lin and Erica Klarreich, 'In times of scarcity, war and peace, a Ukrainian finds the magic in math', Quanta Magazine, 5 July 2022.
227 *most important wave*: James Breiding, *Swiss Made: The Untold Story Behind Switzerland's Success*, Profile Books, 2013, pg 76.
229 *We wanted workers*: Philipp Lutz and Sandra Lavenex, 'Switzerland comes to terms with being a country of immigration', Migration Policy Institute, 18 September 2024.
231 *n'est pas sans rappeler*: John McPhee, *La Place de la Concorde Suisse*, Farrar, Straus and Giroux, 1991, pg 83.
232 *one of the most important*: Kenneth P. Vogel, 'Swiss Billionaire Quietly Becomes Influential Force Among Democrats', *New York Times*, 3 May 2021.
232 *interpret the American Constitution*: Hedi Wyss and Peter Halter, 'Hansjörg Wyss – My Brother', eFeF, 2014.
232 *current generation*: Siddharth Venkataramakrishnan, 'Checkout.com hits $40bn valuation after funding round', *Financial Times*, 12 January 2022.
233 *Switzerland has always been strong*: Oliver Heimes, interview with the author, 10 January 2024.
234 *Switzerland's watch industry*, Federation of the Swiss Watch Industry, 'World Watchmaking Industry in 2023', Annual Report, 2024.
234 *if you think about what has happened*: David Allemann, interview with the author, 29 February 2024.
236 *I would say we were*: Grégoire Ribordy, interview with the author, 15 January 2024.
236 *the single most important*: Anja König, interview with the author, 4 January 2024.

Chapter Seven: The New Mittelstand

238 *Otto Lilienthal*: Markus Raffel and Bernd Lukasch, *The Flying Man*, Springer, 2022.
239 *government plan*: Annie Jacobsen, *Operation Paperclip*, Little, Brown and Company, 2014.
240 *We wanted to prove*: Daniel Wiegand, interview with the author, 7 February 2024.
241 *Lilium has a shot*: Francesco Sciortino, interview with the author, 11 November 2023.
242 *the Kaiser had:* Rafael Laguna, interview with the author, 21 December 2023.
243 *You can do*: Christian Vollmann, interview with the author, 14 December 2023.
246 *transformed the fortunes*: Philip Oltermann, 'Pfizer/BioNTech tax windfall brings Mainz an early Christmas present', *The Guardian*, 27 December 2021.
247 *economic reverberations*: Christiaan Hetzner, 'Pfizer's vaccine didn't just ward off COVID, it may have saved Germany's economy last year', Fortune, 21 January 2022.
248 *concentration of economic power*: Spencer Y. Kwon, Yueran Ma, and Kaspar Zimmermann, '100 Years of Rising Corporate Concentration', *American Economic Review*, Vol. 114, No. 7, July 2024.
250 *hidden champions*: Hermann Simon, 'Lessons from Germany's Midsize Giants', *Harvard Business Review*, March–April 1992.
251 *Hans, don't do this*: Hans Langer, interview with the author, 23 January 2024.
252 *differences are sharp*: André Pahnke and Friederike Welter, 'The German Mittelstand: Antithesis to the Silicon Valley entrepreneurship model?', Working Paper, No. 01/19, Institut für Mittelstandsforschung (IfM) Bonn, 2019.
253 *paper-based accounts*: Stephen Evans, 'Germany's super-shy super-rich', BBC, 28 July 2014.
253 *you'd rather be seen hitting*: Philip Oltermann, 'Germany's Lidl seen overtaking big rivals Tesco, Carrefour, and Aldi', *The Guardian*, 23 June 2014.
254 *Germany's niche companies*: Adrian Wooldridge, 'Germany's niche

companies are a model for life after globalization', Bloomberg, 12 April 2023.
254 *Imagine how much value*: Herbert Mangesius, interview with the author, 6 February 2024.
255 *lack access*: Mai Chi Dao, 'Wealth inequality and private savings: the case of Germany', IMF Working Paper No. 2020/107, 26 June 2020.
255 *A study*: Joe Miller, 'Germany's reclusive rich edge into the limelight', *Financial Times*, 6 September 2020.
256 *rich country*: Chris Bryant, 'Why Germany is rich but Germans are poor and angry', Bloomberg, 15 January 2024.
258 *Rather than thinking*: Sarah Marsh, 'The Mittelstand – one German product that may not be exportable', Reuters, 14 November 2012.
259 *The goal is*: Helmut Schönenberger, interview with the author, January 15, 2024.
260 *In competitive sailing*: Hendrik Brandis, interview with the author, 18 December 2023.
261 *In Europe it's basically impossible*: Dirk Radzinski, interview with the author, 10 January 2024.
262 *It's very simple*: Johannes von Borries, interview with the author, 9 January 2024.
264 *greatest early stage venture bet*: Michael Stothard, 'Early investors in UiPath on track to make a 220,000% return', Sifted, 15 February 2021.
265 *production capacity at home*, Larry Elliott, 'The UK could learn a lot from Germany's long-term industrial strategy', *The Guardian*, 30 March 2016.
265 *creeping de-industrialization*: Matthew Karnitschnig, 'Rust belt on the Rhine', Politico, 13 July 2023.
266 *a little bit behind*: Armin Schmidt, interview with the author, 22 November 2023.
266 *Mechanical outerwear*: Maija Palmer, 'The latest wearable fashion: exoskeletons', *Financial Times*, 17 June 2015.
268 *In the future*: Christian Piechnick, interview with the author, 16 January 2024.

Chapter Eight: Importing Genius

272 *capable of having an original idea*: Melanie Lefkowitz, 'Professor's perception paved the way for AI – 60 years too soon', *Cornell Chronicle*, 25 September 2019.

273 *writing an entire book*: Marvin Minsky and Seymour A. Papert, *Perceptrons: An Introduction to Computational Geometry*, MIT Press, 28 December 1987.

273 *In the eighties*: Garth Gibson, interview with the author, 28 October 2022.

274 *I didn't fail*: Cade Metz, *Genius Makers*, Dutton, 16 March 2021, pg 33.

275 *You may have*: Ibid, pg 30.

277 *Russia!*: Geoffrey Hinton, interview with the author, 13 October 2022.

279 *It was basically invented here*: Jordan Jacobs, interview with the author, 6 October 2022.

280 *We can't sit there*: Ed Clark, interview with the author, 19 October 2022.

282 *Geoff, Yann, and I*: Yoshua Bengio, interview with the author, 5 October 2022.

284 *$32 billion*: National Security Commission on Artificial Intelligence, Final Report, 2021, pg 12.

284 *We're never going to outspend*: Valerie Pisano, interview with the author, 30 September 2022.

285 *It's a good healthy ecosystem*: Sam Ramadori, interview with the author, 29 November 2022.

287 *There are two kinds*: Geoffrey Hinton, 2018 ACM A.M. Turing Lecture, 23 June 2019.

288 *I thought I was going to die*: Richard Sutton, interview with the author, 14 August 2023.

295 *We people*: *The Economist*, 'China needs foreign workers. So why won't it embrace immigration?', 4 May 2023.

295 *demographic character is changing*: Diana Roy and Amelia Cheatham, 'What is Canada's immigration policy?' Council on Foreign Relations, 28 March 2024.

297 *significant revision*: Celia Hatton, 'How Canada's immigration debate soured – and helped seal Trudeau's fate', BBC, 8 January 2025.

299 *The US is not doing itself*: Norbert Lütkenhaus, interview with the author, 2 November 2022.

300 *60 per cent of the respondents*: 'The Student Voice: National Results of the 2023 CBIE International Student Survey', Canadian Bureau of International Education, 2024.
301 *I just think*: Damien Steel, interview with the author, 11 October 2022.
302 *I have absolutely no doubt*: Sanja Fidler, interview with the author, 31 August 2023.

Conclusion: Social Animals

304 *Look at any measure*: Adrian Wooldridge and Alan Greenspan, *Capitalism in America*, Allen Lane, 2018, pg 392.
310 *I don't think*: David Allemann, interview with the author, 29 February 2024.
312 *Tech hubs are losing*: Christopher Mims, 'Tech hubs are losing the talent war to everywhere else', *Wall Street Journal*, 22 December 2023.
312 *If you ski*: Nathalie Casas, interview with the author, 8 February 2024.
313 *Every café*: John Kim, interview with the author, 25 October 2022.
313 *a fifth:* Dealroom, 'Central and Eastern European Startups', November 2022.
313 *Innovation is the child*: Matthew Ridley, *How Innovation Works: And Why It Flourishes In Freedom*, HarperCollins, 19 May 2020.

Index

Page entries in *italics* indicate images.

01.AI (*Ling-Yi Wan-Wu*) 54–5
3D printing 250–51
3M 249
5G 8, 42, 46, 191, 192
12038 138–9

Accel 101
Activision Blizzard 20–21
Aebischer, Patrick 217–21, 225, 235
Agarwal, Ajay 3–5
Ahn, Ikkjin 168
AI. *See* artificial intelligence
AI Foundation Models Taskforce, UK Government 100
Airbnb 51–2, 95, 117
Aixtron 25
Alberta Machine Intelligence Institute (Amii) 282, 283, 284, 286, 291, 301
Albrecht brothers 252–3
Aldi 252–4
Aleph Alpha 259, 261, 262
AlexNet ('ImageNet Classification with Deep Convolutional Neural Networks') (Hinton) 32, 276, 278, 291, 292
Alibaba 17, 25, 33, 38, 52, 97, 140, 172, 178, 179, 190, 191, 193
Allemann, David 234–5, 310
Alphabet 64, 74, 102, 121, 136
AlphaGo 31, 98, 287
Alphawatch 39
AlphaZero 31, 287
Alternative for Germany (AfD) 294
Altos 166, 220
Amadeus 27

Amazon (technology company) 18, 45, 51–2, 68, 74, 139, 140, 172, 178, 248
Amodei, Dario 74
Anbang 23
Annus, Toivo 170
Ant Group 52
Anthropic 66, 74, 100
Antina 192
Apollo autonomous ride hailing service 43
Apollo space program, US 88, 239
Apple 6, 8, 9, 14, 25, 65, 66, 74, 81, 84, 89, 95, 102, 123, 142, 150, 151, 188, 190, 208, 223, 224, 241, 245, 265, 268, 298, 305
Applied Machine Learning Days (AMLD) 222
Arm 8, 27, 115, 123–8, 306
artificial intelligence (AI) 7, 26; AlexNet paper 32, 276, 278, 291, 292; Canada and 212, 270–92, 297, 301–3; China and 30–39, 45, 54–5, 66–75, 284, 302; defined 30, 271; Germany and 259, 262; ImageNet competition 31, 32, 276, 278; machine learning 5, 7, 30, 38, 73, 109, 222, 276, 278–9, 283, 289; neural networks 30–32, 42, 271–3, 275, 276, 286–7, 299; origins 270–75; perceptron and 272–3, 275; Reinforcement Learning, or RL and 286–90; ResNet paper ('Deep Residual Learning for Image Recognition') 30–33, 35, 37, 40; South Korea and 146, 163, 164; symbolic or rule-based AI (Good Old Fashioned AI, or GOFAI) 271–2; UK and 14, 31, 98–101, 106, 108–9, 115, 119, 120, 123–4, 127–30, 135–6, 137, 287, 302, 310; US and 10, 38, 54,

Index

66–77, 100, 102, 109, 123, 125, 126, 127, 129, 130–31, 146, 188, 198, 245, 259, 261, 262, 272, 276, 278, 290, 302, 310; Vietnam and 188. *See also individual company name*
ASML 8
Association of Southeast Asian Nations (ASEAN) 172, 185–91
Atomico *11*, 108
autonomous vehicles 31, 42–3, 45, 242, 276
Aveva 127
aviation 88, 238–41

B-70 bomber 88
B Capital 173
Baden-Württemberg, Germany 250, 253, 259
Baer, Hans 228
BAIC Group 23
Baidu 33, 35, 43, 45, 46, 67, 72–4, 97, 179
Balakrishnan, Vivian 198
Blavatnik, Leonard 231
Balls, Ed 112
Bankman-Fried, Sam 61
Barber, David 119–20
Barlow Commission Report (1940) 113–14
Barton, Dominic 293
Basel, Switzerland 219, 220, 236
BASF 243, 244, 265, 312
Benaich, Nathan 73, 121–2
Bengio, Yoshua 212, 270, 276, 282–3, 284, 286, 291, 292
Benz, Karl 243
Berner, Bertha 78–9
Berners-Lee, Tim 213–14
Berset, Alain 221
Bessmer 96
Biden, Joe 24
Bigo Live 191
billion-dollar tech companies, top ten countries by number of private *9*
BioNTech 246–7
Bismarck, Otto von 242
bitcoin 61–2, 115–16
BlackRock 141, 293
Blackwell, Theo 105, 107, 202
blitz scaling 111, 135
blockchain 61–3, 115–17, 227
Bloomberg Innovation Index 204
Blue Brain 210
BlueTrace 223
BMW 243, 244, 248, 259, 265, 267, 312

Boltzmann Machine 275
Boole, George 274
Boosfeld, Maximilian 216
Bordier 228
Boring Company 250, 257–8
Borries, Johannes von 262
Boston Consulting Group 138
BrainBox AI 285
brand advantage 233–4
Brandis, Henrik 260
Braun, Karl Ferdinand 243
Braun, Wernher von 239
Breiding, James: *Swiss Made* 228, 234
Broadcom 69, 188, 222
Brudermüller, Martin 265
BTC-e 63
Bugnion, Edouard 69–70, 222, 223, 235
Burke, Dave 224
Burkhalter, Didier 221
Bush, Vannevar 84–7
Buterin, Vitalik 227
BYD 8, 27, 48, 51, 167, 193, 265
ByteDance 8, 14, 33, 100, 172, 178, 191, 302

C1 243
Calico 220
California, US 17, 29, 34, 37, 40, 49, 57, 79, 80, 82, 83, 86, 91–2, 105, 262, 275, 287, 292. *See also* Silicon Valley
Californian, SS 83
CalPERS (California Public Employees' Retirement System) 262
Caltech 34, 37, 40, 86
Cambridge, UK 27, 109, 112, 113, 114, 123, 125, 198, 226, 274
Canada 5, 21, 127, 212, 226, 270–303; AI and 212, 270–92, 297, 301–3; Alberta Machine Intelligence Institute, or Amii *see* Alberta Machine Intelligence Institute (Amii); Alberta Plan for AI Research 291; Bengio and *see* Bengio, Yoshua; Canadian Institute for Advanced Research (CIFAR) 283–4, 302–3; Hinton and *see* Hinton, Geoffrey; immigration and 292–300, *296*; international students 300; Montreal Institute for Learning Algorithms, or Mila *see* Montreal Institute for Learning Algorithms (Mila); Pan-Canadian AI Strategy 212, 282–4 *see also individual institute name*; Sutton and *see* Sutton, Rich; Tech Talent Strategy 299;

Canada (cont'd)
    top ten countries by number of private billion-dollar tech companies and 9; Vector Institute for Artificial Intelligence *see* Vector Institute for Artificial Intelligence; venture capital in 10, 279, 300–301, 309
Canadian Natural Sciences and Engineering Research Council (NSERC) 290
capital 20, 22–5, 27, 28, 39, 53, 181, 183, 236, 241, 256, 257, 261, 262, 301, 312; venture capital *see* venture capital
capitalism 7, 18, 54, 58, 121, 125, 134, 135, 160, 255, 310
Capsitec 246
capsule networks 299
car industry 23, 28, 46, 146, 209, 230–31, 242–4, 248, 259, 265, 267, 312, 316–17; autonomous vehicles 31, 42–3, 45, 242, 276; electric vehicles (EV sector) 27, 28, 42, 50, 146, 187, 191, 209, 240
caravan society 103
Carnegie, Andrew 81
Carousell 184–5
Casas, Nathalie 217, 312
CATL 8, 187
Celonis 260, 266
Century Initiative 293
CERN (European Organization for Nuclear Research) 208, 211–14, 312
Chainalysis 61–3
ChatGLM 38
ChatGPT 38, 272, 276, 290
Checkout.com 101, 232
Chen, David 176
Chevrolet 230–31
China 16–55; artificial intelligence and 30–39, 45, 54–5, 66–75, 284, 302; autonomous driving technology and 42–3, 45; billion-dollar tech companies, number of private 9, 101, 102; Canada and 278, 279, 284, 295, 296, 299, 300, 303; chip imports 25–6, 45, 68–9, 73, 128; Circular Electron Positron Collider (CEPC) 214; clean energy production 205–6; collaboration with US in tech research 30–40; competence in tech 7–8; consumer, adoptive nature of 48–9; electric vehicles (EV sector) 8, 27, 28, 42, 48, 50, 51, 167, 193, 265; freedom in, innovation and 313–15; Germany and 248, 252, 256, 257, 258, 259, 260, 262, 265–6, 268; global GDP, contribution to 307; global outbound investment 23–4; government crackdowns on tech 28–9, 52–4; government, public support for 49–50; government subsidies, tax incentives and direct investments 49; immigration and 47, 295, 296, 296, 299, 300; implementation, agility in 45–7; Initial Public Offering (IPO) process 50–51; Nobel Prize and 40–41; Singapore and 169, 170, 174, 176, 177, 180, 181, 186–93; South Korea and 140, 141, 143, 144, 165;
    Special Economic Zones 43; state-owned enterprises 23, 51, 308–9; Switzerland and 205–6, 210, 214, 218–19, 223, 233, 234; Tencent *see* Tencent; Total Tech Market Cap ($T) per region (2021–2023) and 11; universities 31, 33, 34–41; UK and 101, 102, 103, 114, 127, 130, 131, 134, 135; US and *see* US; venture capital and 10, 22, 23, 24, 25, 26, 27, 38, 42, 52, 54, 97; work ethic in 314.
    See also individual place name
China Construction Bank 308–9
China Initiative 73–4
Chinese Academy of Sciences (CAS) 37
Chinese Communist Party (CCP) 24, 33, 36, 53
Cho, Heather 155
Choi Soon-sil 153–4
Christensen, Clayton 139, 150
Christensen, Matthew 138, 139
Chung-hee, Park 144–5, 153
Circular Electron Positron Collider (CEPC) 214
citadel society 103
Clarivate 32
Clark, Ed 280–82, 293, 296–7
Clash of Clans 2–3, 20–21
Clifford, Matthew 120
clusters 66, 69, 91–2, 105, 114, 167, 259, 311
Coates, Adam 74
cognitive robots, or cobots 26
Cold War (1946–91) 75, 90, 143, 165, 166, 188, 189, 192
Collison, Patrick 72
colocation 110–12, 313
'common prosperity' 53, 54
compensation 77, 89, 90, 103–4, 199

Index

compute (processing power) 67, 69–70, 129, 130, 289–90
Cook, Tim 188
Coupang 139–41, 150, 163, 165, 166
Covariant 287–8
Covid-19 223, 246
Creative Destruction Lab (CDL) 4–5, 301
Creative Technologies 180
Crick Institute 106
CRISPR technology 167
cryptocurrency 52, 61–3, 115, 116, 117, 227
CSIS 50
Cultural Revolution (1966–76) 34
CureVac 247

Daewoo 156
Daimler, Gottlieb 243
DANA 191
Dapper Labs 292
Darktrace 127
Dartmouth College 86, 270–71
Databricks 66, 100
DCM Ventures 24
Decentralized Privacy-Preserving Proximity Tracing (DP-3T) 223–5
decisions intelligence 109–11
Deep Genomics 5, 302
DeepMind 14, 31, 98–101, 106, 109, 115, 120, 123–4, 127, 135–6, 287, 302, 310; Ethics and Safety Review Agreement 136
DeepMind Labs Limited 135
Deep Speech 2 73, 74
deep tech 27, 99, 181, 195, 215, 232, 238, 242, 250, 251
Defense Advanced Research Projects Agency (DARPA) 88, 97, 120, 212, 245, 247, 284, 309
Delangue, Clement 7
Deng Xiaoping 43, 48, 53, 192
Depop 127
DeskOver 263–4
Didi Chuxing 25, 52, 190
Dines, Daniel 264
direct investment 23, 50
Distributed Ledger Technologies (DLTs) 116–17
DIW 255
DJI 8, 48
DNA origami 246
DNNresearch 278
Doerr, John 96

Dorsey, Jack 310
Dota-2 289
Doudna, Jennifer 167
Durant, William 230, 231
Dychtwald, Zak 48

Earlybird 260–64
eBay 72, 139, 172
École Polytechnique Fédérale de Lausanne (EPFL) 119, 217–18, 221–3, 225, 229, 232, 235
Einstein, Albert 216, 227
electric vehicles (EVs) 27, 28, 42, 50, 146, 187, 191, 209, 240
Electro Optical Systems (EOS) 250–52
Electronic Arts 2
Ellison, Larry 60
emoji 7
Epic Games 20
Ericsson 150, 192
Ethereum 116, 227
Etsy 61, 127
European Commission 26, 204
European Investment Fund (EIF) 237
European Nasdaq concept 236
European Policy Analysis Group 13
European Union (EU) 13, 115, 128, 186, 191, 229–30, 236–7
eVTOLs, or electric Vertical Take off and Landing Aircraft 240
Exeter boarding school, New Hampshire 163–4
exoskeletons 266
Explorer-1 239
exports 68–9, 81, 131, 134, 142, 145, 146, 148, 162, 186, 187, 219, 234, 244, 248, 265, 280

Facebook 9, 18, 19, 66, 72, 95, 106, 110, 161, 173, 197, 198, 254, 276
facial recognition 31, 33, 34, 49
Faculty 108–9
Fairchild 88–9
Farshchi, Shahin 24
FBI 73
FOMO (fear of missing out) 95–6
Federal Telegraph Company (FTC), US 83–4
Federal Trade Commission (FTC), US 126
Federer, Roger 203–4, 226–7, 234
Fidler, Sanja 302
Fields Medal 37, 225, 226
financial crisis (2008) 170

Financial Times 57, 121, 124, 132, 133
Flixbus 260
Fortnite 20, 171, 172
Fortune 500 48, 51, 207, 233
Franco-Prussian War (1870–71) 242
Free Fire 171–2, 174
freedom and prosperity, innovation and 313–15
Frisch, Max 229
FTX 61
Fujian Grand Chip Investment Fund 25
Funcom 21
Future Circular Collider (FCC) 214
Future Life Church 153–4

GAEN 223
gaming 1–3, 8, 20–21, 25, 28–9, 52, 68, 115, 161, 165, 166, 170–73, 175
Gang Chen 73
Gang Ye 176
Garena 170–71, 173
GDP 48, 77, 112–13, 121, 147, 148, 171, 185–6, 205, 215, 218, 242, 247, 248, 305, 307
Geely 23
Geim, Andrej 117–19
GenEdit 167
GeneEdit 163
Genentech 218, 219
General Motors (GM) 146, 230, 231, 244
Geneva Conventions 229
German Bionic 266–7
Germany 22, 48, 71, 84–5, 101, 112, 127, 186, 223, 225, 233, 238–69, 312, 314; aviation industry 238–42; billion-dollar tech companies *9*; biotech industry 245–7; capital availability 241, 256–62; car industry 240, 242, 243, 244, 248, 259, 265, 267; China and 25–6, 248, 252, 256, 257, 258, 259, 260, 262, 265, 266, 268; de-industrialization in 265–6; economy, size of 247–8; European wealth rankings and *257*; exports 244, 248, 265; Federal Agency for Disruptive Innovative (SPRIND) 242, 245–7; *Gründerzeit*, or Founder's Era 242–3, 245, 265; 'hidden champions' 250–51; immigration and 246, 294, 295, 296, 299, 300; incremental optimization 249, 258; Industry 4.0 strategy 266, 267; manufacturing base 260, 264–5; Mittelstand 248–61, 308; new technologies, problems building and commercializing 243–5; Operation Paperclip and 239; origins of 242; population numbers 248; risk, cultural aversion to 261–2; robotics and 242, 246, 263–4, 266–9; venture capitalism and *10*, 245, 254, 259–61, 262; wealth inequality in 254–6, *255*; WIPO Global Innovation Index and *12*
GG 170
GGV Capital 24
Gharegozlou, Roham 292
Gibson, Garth 273–5
GIC 184, 193
Giving Pledge 160
Global Innovation Index (GII) 11–12, *12*, 203
Global Switch 101
Go 31, 287
golden shares 54
Goldman Sachs 253
Google 6, 9, 37, 41, 45, 46, 64, 65, 66, 68, 72, 74, 95, 96, 99–100, 106, 110, 121, 123, 128, 132–3, 135, 136, 158, 162, 178, 195, 197, 199, 200, 201, 208, 210–11, 218, 223, 224, 229, 241, 245, 265, 278, 282–3, 299; DeepMind and *see* DeepMind; Google Maps 110, 161, 229
Gopher 213
Gothard Base Tunnel 209–10
Grab 182–4, 190, 306, 309
Granary Building 105–6
Graphcore 128–31
graphene 117–19
Graphical Processing Units (GPUs) 68, 70, 129
Greenspan, Alan: *Capitalism in America* 304
Griffin, Ken 231
Grinding Gear Games 21
Gronager, Michael 63
Groupon 139
Gunshine 2
Gutenberg, Johannes 247

Hadfield, Chris 5
Han's Group 26
Hanjin Group 155
Hanton, Angus: *Vassal State* 127
Harvard Business School 138, 150, 182, 231
Harvard College 6, 37, 39, 50, 85–6, 87, 90, 91, 108–9, 138–41, 159, 216–17, 220, 225, 231, 251, 279, 280

Index

Hassabis, Demis 120, 136
Hauser, Hermann 27–8, 124, 126, 129
Hawkins, Trip 2
Hay Day 2
Haye's Valley 66–7
Hayek, Nicolas 226, 228
Heimes, Oliver 233
Heinkel He 178 239
Hennesy, John 64–5, 76
Herrenknecht 250, 252, 256–8
Hewlett-Packard 87–8
hidden champions 250–51
Hillhouse Capital 191
Hinton, Geoffrey 31–2, 270, 273–9, 282–3, 286–9, 291, 292, 296, 299; AlexNet ('ImageNet Classification with Deep Convolutional Neural Networks') 32, 276, 278, 291, 292
Hitachi 93
Hofmann, Albert 220
Hogarth, Ian 100
Hu Jintao 36
Huang, Jensen 188
Huawei 23, 48, 51, 191, 192
hubs 7, 61, 90, 93, 105, 107, 114, 182, 215, 227, 229, 235, 303, 311–13
Hugging Face 6–7, 306
Hughes, Dan 115–17
Huguenots, French 227–8
Human Brain Project 213
Hunt, Jeremy 121
Hwang, Victor 58–60; *The Rainforest* 59–60
HYBE 161
hyper scaling 111
Hyundai 142, 145, 146, 152, 153, 162, 187, 267

ICQ 16
ID Quantique 235
iFlyTek 34
ImageNet computer vision contest 31–2, 276, 278
IMD World Talent Ranking 204
immigration 41, 47, 72, 103, 170, 173, 175, 220–21, 225–30, 246, 292–300
incremental optimization 249, 258
Index 108, 132–4
India *9, 10*, 44, 141, 165, 172, 174, 185, 186, 201, *296*, 300, 306, 309
Indonesia 172, 177, 180, 184, 185–7, 191
Inflexion 21
info.cern.ch 213

Initial Public Offering (IPO) process 51, 52, 139, 140, 141, 173, 205, 254, 263–4
INSEAD Global Talent Competitiveness Index 204
Institute for Quantum Computing 299
Institute of Electrical and Electronics Engineers (IEEE) 35
Intel 89, 95, 129, 150–51, 188
intellectual property rights 19, 27, 29, 123, 124, 180, 213, 252
Intelligence Processing Units (IPUs) 129
International Committee of the Red Cross (ICRC) 228–9
International Data Group (IDG) 23–4
International Monetary Fund (IMF) 156, 255; 'Changing Global Linkages: A New Cold War' 189
International Olympic Committee (IOC) 229
International Space Station (ISS) 5, 211
International Thermonuclear Experimental Reactor (ITER) 212–13
involution 53
iPad 151
Isar Aerospace 261
Israel *9*, 22, 122
Iswaran, Subramaniam 192

Jacobs, Jordan 279, 281, 298–9, 303
Japan 20, 32, 71, 84, 86, 93, 123, 125, 126, 127, 131, 140, 141, 142–3, 144, 147, 149 150, 159–60, 177, 186, 191, 233, 241, 245, 266, 304, 308, 314–15
jet propulsion 239
Jian Sun 31
Jiang Zemin 43
Jie Tang 38–40
Jobs, Steve 14, 89, 139, 150, 151, 169, 174
Johnson, Boris 113
Johnson, Elsbeth 50–51
Joint European Disruptive Initiative (JEDI) 212
JP Morgan 253
Julius Baer 228
Jupiter-C 239

Kaiming He 31
KakaoTalk 158–62
Kanji, Hussein 26, 103
Kano 110
Karp, Alex 60

Kauffman Fellows Program 59
Kepler 5
Keynes, John Maynard 13, 48
Khazanah Nasional 183
Khosla Ventures 24, 60
Khosrowshahi, Dara 183
Kim Beom-su 158–62
Kim, Bom 138–43, 165
Kim Dae-jung 157–8
Kim, Han 165–6
Kim, Jimmy 157
Kim, John 164–5, 312–13
King's Cross, London 105–8, 113, 114, 178
Klamer, Arjo 103
Klein, Alex 110
Klein, Saul 107–8, 115, 132–3
Kleiner Perkins 96
Knowledge Quarter, London 106
Kodisoja, Mikko 1, 2
König, Anja 236–7
Korea Advanced Institute for Science (KAIST) 76, 167
Korean Air 155
Krafton 21, 163, 166
Kraken 63
Krizhevsky, Alex 278, 292
Kuka 25, 267

L-shaped returns 94–5
Laguna, Rafael 242, 244, 245–7
Lakestar 233
Langer, Hans 251–2
Large Language Models (LLMs) 38, 42, 390
Lazada 172, 190, 191
League of Legends 20, 170
LeCun, Yann 276, 282–3
Lee Byung-chul 145, 148, 151
Lee Hae-jin 158
Lee Hsien Loong 193, 197, 198, 200
Lee, Jae 163–4
Lee Jae-yong 154
Lee, Jay Y. 160
Lee Jong-beom 156
Lee, Kaifu 41–2, 44, 54–5, 310, 314
Lee Kuan Yew 175–6, 177, 179, 192–3, 200
Lee Kun-hee 149, 153
Lee, Kunwoo 167
Lee Myung-bak 152, 153
Leiden Ranking 36
Leland Stanford Junior University 80
Lens Technology 190

Leuthard, Doris 221
LG 8, 142, 145, 187
Li Shipeng 34, 37
Li Yang 53
Li, Hongyi 199
Li, Xiaodong 'Forrest' 169–75
Lilienthal Normalsegelapparat 238
Lilienthal, Otto 238–40
Lilium 26–27, 239–42, 261
Line 159–60
Ling, Tan Hooi 182
Lived Change Index 48
LocalGlobe 133
LOGIbodies 246
Lombard Odier 228
London. *See* UK
LSD 220
Lunit 163
Lütke, Tobias 292
Lütkenhaus, Norbert 299–300
Lux Capital 24, 74
Luxshare Precision Industry Co 190
Lydia 25

Ma Huateng 16, 17
Ma, Rui 39, 49, 50
machine learning 5, 7, 30, 38, 73, 109, 222, 276, 278–9, 283, 289
Macron, Emmanuel 6, 26
Mao Zedong 34, 53, 192
Mainz, Germany 246–7
Malaysia 172, 177, 182–7, 191
Mallaby, Sebastien: *The Power Law: Venture Capital and the Making of the New Future* 93–4, 96, 97
Mangesius, Herbet 254
Manhattan Project 84, 100, 274
manufacturing 68, 97, 114, 115, 136, 146, 162, 167–8, 187, 188, 190, 258, 260, 264–5
Marginson, Simon 36, 41, 50
Maschinenfabrik Otto Lilienthal 238
Max Planck Institute for Plasma Physics 251
McGill University 283, 301
McPhee, John: *La Place de la Concorde Suisse* 231
Megvii 33, 37
Meituan 191
Menlo Park, Palo Alto 83
Mercedes-Benz 23, 243, 265
Merrill Lynch 280–81
Messerschmitt Me 262 239

Meta 68, 74, 102, 178, 283
Miami, US 61, 114
microchips 8, 10, 25, 45, 67–70, 73, 74, 76, 102, 123, 125, 126, 128–32, 146, 151, 168, 187, 188, 265, 302
Micron 188
Microsoft 45–6, 65, 67, 68, 72, 74, 102, 121–2, 142, 178, 188, 254, 268, 282; Microsoft China 35; Microsoft Research Asia (MSRA) lab, Beijing 30–31, 33–5, 37, 41–2; Microsoft Research China 32
Midea Group 25, 267
Minsky, Marvin 273, 275–7, 287
Mirabaud 228
MIT (Massachusetts Institute of Technology) 33, 36, 37, 38, 40, 73, 84, 86, 87, 90, 91, 119, 199, 216, 273, 275, 279
Mitsubishi 149, 308
Mittelstand 248–61, 308
Mitter, Rana 50–51
Moderna 247
Moloco 163, 168
Montreal Institute for Learning Algorithms (Mila) 282–6, 297, 300, 301
Moore, Gordon 89
Moritz, Michael 57
Motorola 93, 150
mRNA, or messenger RNA 246–8
Mt. Gox 61–3
Mulroney, Brian 280, 281
Musk, Elon 60, 109, 146, 250
Mynt 191
MyTeksi 183

N26 25
Nadella, Satya 188
Naipaul, V. S. v
nanorobots 246
Naspers 16–18
National Aeronautics and Space Administration (NASA) 88
National University of Singapore 182
Nature Index 32, 36
Naver 158–60, 162
NEC 93
Nespresso 220
Nestlé 218, 226, 233, 236
Neumann, John von 216
neural networks 30–32, 42, 271–3, 275, 276, 286–7, 299
Neura-robotics 25–6

New Palo Alto 108, 135
New York, US 61, 112, 114, 126, 141, 155, 204, 206, 218, 221, 229, 249, 250, 309
New York Stock Exchange (NYSE) 140, 173, 205, 263
NeXT computer 213
NFC 115
Ng, Andrew 72–4
Niel, Xavier 6
Ninja Van 184–5, 190–91
Nio 191
Nobel Prize 13, 32, 88, 119, 167, 218; China and 40–41; Germany and 245; Switzerland and 207, 215, 225; US and 40; winners per capita 215
Nokia 3, 111, 150, 168, 192
Nortel Networks 168, 301
Novartis 218–20, 233, 236
Novoselov, Konstantin 117–19
Nvidia 10, 67–70, 74, 102, 129, 130–31, 146, 188, 265, 302; A100 68; Arm takeover bid 123, 125, 126, 127; Graphical Processing Units, or GPUs 68, 70, 129; H100 68, 129

O'Mara, Margaret 89–90
Obama, Barack 50, 189
Office of Naval Research, US 87
Office of Scientific Research and Development (OSRD), US 84
OMERS Ventures 301
ON 234, 310
OpenAI 38, 54, 66, 74, 100, 109, 198, 245, 259, 261, 262, 278, 290, 302, 310
OpenAI Five 289

Paananen, Ilkka 1, 28
Palo Alto, US 7, 17, 22, 65, 66, 69, 78, 80, 81, 83, 107–8, 135, 169, 207, 263
Pan-European Privacy-Preserving Proximity Tracing (PEPP–PT) 223
Paperclip, Operation 239
Park Geun-hye 153–4
Paul Scherrer Institute (PSI) 214–15
Paulson Institute 33
PayPal 95, 104, 116, 173
Penang 187–8
pension funds 262
Pentagon, US 24, 136
Pfizer 246
pharma industry 25, 218–20, 228, 233, 242, 246

Philips 93
Pictet 228
Piechnick, Christian 268–9
Piketty, Thomas 53
Pisano, Valerie 284, 285, 297, 300
Pixar 14
Plectonic 246
Poilievre, Pierre 294
Pousaz, Guillaume 230, 232
product lifecycle theory 91
Project Maven 136
Project Mercury 239
PropertyGuru 184–5
Proxima 119, 241
PUBG 21, 142, 163, 171, 172
Pudong 43–4

Qiushi 53
QQ 16, 17, 18, 20
QR codes 49
quality of life 130, 221
quantum technology/quantum computing 5, 214, 216, 235–6, 283, 292
Quibi 253

R&D (research and development) 12, 14, 38, 84, 86, 87, 88, 117, 153, 180, 188, 212, 215, 245, 284
Rabois, Keith 24, 25, 60–61, 66
Radical Ventures 279
Radio Research Lab (RRL), Harvard 85, 87
Radix 115, 116
Radzinski, Dirk 261
rail networks 45, 107–8, 208–11
Ramadori, Sam 285
Raspberry Pi 112
reactive vulnerability 236
Reagan, Ronald 71
recycling of talent 158, 222–3
Reddit 22, 32, 115, 131, 165, 313
Redstone 239
Reduced Instruction Set Computer (RISC) 76
Regional Comprehensive Economic Partnership (RCEP) 186
Reinforcement Learning (RL) 286–90
Research in Motion (RIM) 301
ResNet paper ('Deep Residual Learning for Image Recognition') 30–33, 35, 37, 40
Revolut 101, 115
Ribordy, Grégoire 235–6

Ridley, Matt: *How Innovation Works: And Why It Flourishes in Freedom* 313
Rimer, Neil 133
Riot Games 20, 170
risk, attitudes towards 39, 94–7, 99, 103, 120, 156, 241, 243–4, 260–63, 284, 306
Roblox 166
robotaxis 43, 46
Robotic Process Automation (RPA) 266
robotics 25–6, 42, 178, 216, 233, 242, 246, 263–4, 266–9, 287, 304–5
Roche 219, 226, 233
Roosevelt, Theodore 35, 84
Rose Park Advisors 139, 140
Rosenblatt, Frank 272, 273
Route 128, US 91
Rüegg, Christian 214, 215
Ruse, Konrad 243
Rycker, Sonali de 101, 108

Sabour, Sara 249
Sacks, David 121–2
Saich, Tony 50
Salathé, Marcel 222–5
Samsung 8, 106, 142, 145, 147–54, 157–60, 162, 167, 180, 187; Samsung Biologics 167–8; Samsung Group 147
San Francisco, US 6, 7, 46, 56–8, 60–61, 63, 66–70, 74, 76–8, 80, 83, 90–93, 105, 107–8, 110, 114, 125, 133, 135, 164, 204, 206–7, 219, 249; Bay Area 6, 7, 56–7, 60–61, 66, 67, 69, 70, 74, 76–8, 83, 90–93, 105, 107–8, 110, 135, 204, 206–7, 219
sanctions, government 24, 37, 186, 190
Sandoz laboratories, Basel 220
Saturn-V 239
Saverin, Eduardo 173
Saxenian, AnnaLee 90–93; *Regional Advantage: Culture and Competition in Silicon Valley and Route 128* 91
Schmidt, Armin 266–7
Schmidt, Eric 96, 284
Schneider 127
Schönenberger, Helmut 259–60
Schwarz Group 259
Schwarz, Dieter 253
Sciortino, Francesco 119, 241, 242
Sea 173–6, 178, 184, 190
SeaMoney 172–3
Second World War (1939–45) 84–7, 220, 239

Index 353

seed round 104
Segars, Simon 124–5
semiconductors 8, 25, 66, 76, 88–90, 93–4, 114, 118, 119, 123, 130–32, 151, 168, 187–8
Sendbird 163, 165, 312–13
SenseTime 34
Sequoia Capital 24, 57
SG Tech initiative 181
Shaoqing Ren 31
Sharkmob 21
Shein 191, 306
Shenzhen, China 16, 43, 47–8, 267, 310
Shenzhen-Zhongshan Link, China 47
Shing-Tung Yao 37
Shockley Semiconductor Laboratory 88
Shockley, William 88
Shopee 172, 173, 190
Shopify 292, 301
SICPA 235
Siemens 232, 243, 248
silicon 45, 78, 118
Silicon Valley, California 4, 5, 8–9, 310, 311; Canada and 283, 301, 305; China and 22, 23, 33; Germany and 238, 251, 252, 259; high-profile departures of major tech figures 60–61; legacy 60; London and 99, 105, 109, 110, 111, 120, 130, 132, 133, 134, 135, 137; non-compete clauses and 91–2; origins 77–90; per capita GDP 77; premature obituaries for 56, 57, 74–5; Route 128 and 91; Singapore and 197, 201; success of, reasons for 90–97; Switzerland and 207, 222
Simon, Hermann 250
Singapore 69, 113, 141, 169–202, 223, 224, 232, 233, 266, 292, 306–7, 309; ASEAN and 185–91; automated guided vehicle (AGVs) in 197; Block71 178; bureaucratic control in 178–9; Changi airport 180; China and 169, 170, 172, 174, 176, 177, 180, 181, 186–7, 189–93, 201; data.gov.sg 196; Digital Government Exchange (DGX) 197, 198; Economic Development Board 178; government grants/aid to business 181–5; government lays foundations for high-tech economy 175–6; Government Technology Agency (GovTech) 195–9; Grab and 182–4, 190, 306, 309; immigration and 175–7; independence, gains 176–7; National Computerization Program 195; One-North neighborhood 178; public authorities use of new technologies 193–6; public sector, channels best talent into 197–8; R&D capacity of indigenous firms 180; regional market and 184–91; Singpass 196; smart cities and 193–202; Smart Nation and Digital Government Group, or SNDGG 195; Smart Nation and Digital Government Office, or SNDGO 195; Smart Nation Fellowship 198; Smart Nation initiative 195–6; Smart Nation Sensor Platform 196; Strategic National Projects, or SNPs 196; tax exemptions in 181; Technopreneurship Innovation Fund, or TIF 179; Tuas Port 197; venture capital in *10*, 173, 179–82; Virtual Singapore 196; WIPO Global Innovation Index 12
Singtel 192
Sinochem 23
Sinopec 308–9
Sinovation 42
Sixteen Thirty Fund 232
SK Hynix 146, 147, 187
SK On 8
skip connections 30–31
Skype 104, 108, 125, 170
SM Entertainment 161
smart cities 193, 202
Smart Cities Index 193
Snap 22, 95, 127
Softbank 17, 123, 131, 141
Son, Masayoshi 123–4
Song Young-gil 161
South Korea 8, *12*, 20, 21, 76, 139–68, 186, 187, 233, 305; Asian Financial Crisis and 155–7; Candlelight Revolution 154; chaebols 145–8, 151–62, 167–8, 309 *see also individual company name*; Coupang 139–41, 150, 163, 165, 166; culture of impunity for powerful in 153–5; digital economy away from chaebol control, origins of 157–62; election (1997) 157; *gabjil* (high handedness) 155; growth curve 142–5; hybrid companies with presence in US and Korea 163–8; Korean War 155–6; Kosdaq 158; origins of 142–4; Samsung and *see* Samsung; third generation of tech companies 162–8; venture capital in *10*, 157–8, 166; WIPO Global Innovation Index and *12*
SPAC 183
SpaceX 72, 240, 251, 261, 267

SparkLab 157
Spotify 8, 21, 205, 306, 309
Sputnik 75, 88
Square Mile, London 105
St Lucia 215
Stanford University 14, 31, 38, 40, 54, 65, 69, 72, 76, 90, 96, 166, 169, 170, 174, 176, 215, 216, 218, 222, 276, 279, 288; Applied Electronics Lab 87; origins of 76–88; Silicon Valley origins and 94; Stanford Industrial Park 88; Stanford Research Park 215
Stanford Junior, Leland 78–82
Stanford, Jane 79, 80
startup culture 3, 6, 9, *9*, 10, 14, 16, 17, 18, 89, 242; Canada 278, 279, 283, 285, 297, 300–301, 303; China 16, 17, 18, 24–5, 27, 33, 38, 39, 45, 48, 51, 54; Germany 241–2, 243, 252, 259–63, 266; Singapore 138–40, 157, 158, 163, 164, 166, 170–71, 178–84, 188, 190, 193, 195; South Korea 305, 309, 311, 312–13; Switzerland 215, 218, 232, 233, 234, 235, 236; UK 98–101, 103, 108–10, 115, 119, 125, 127, 128, 132, 133; US 61, 63, 68, 69, 72, 87, 89, 96
Startup Genome 181
Startup SG Founder 181
Station F 6–7, 178
Steel, Damien 301, 302
Sternbach, Leo 226
stock markets 51–2, 74, 97, 102, 104, 121, 140, 146, 158–9, 161, 163, 166, 173, 174, 205, 235–6, 240–41, 252, 254–6, 262–4, 268
Stripe 72, 98–9, 232, 309
subsidies 50, 145
Sumea 1–2
Sumo Group 21
Sun Microsystems 72
Supercell 2–3, 20, 28
superorganisms 316
Sutskever, Ilya 278, 292
Sutton, Rich 270, 286, 288–92
Swatch 226, 228
Swiss Federal Institute for Technology (ETH) 69, 215–18, 222, 223, 226, 229, 231
Swiss Federal Laboratories for Material Science and Technology 217, 312
Swiss National Railways 211
Swiss Patent Office 227
Switzerland 69, 82, 113, 119, 127, 133, 203–37, 250, 298, 310, 312; Blue Brain project 210; brand advantage 233–4; breaks from consensus 207–8; Covid-19 and 223; Crypto Valley 227; Decentralized Privacy-Preserving Proximity Tracing (DP-3T) 223–5; diaspora 230–33; European Investment Fund (EIF) 237; Federal Council 207, 221; finance in 210, 228, 232, 236; foreign academics working in universities 225; Fortune 500 companies 233; French Huguenots in 227–8; Future Circular Collider (FCC) 214; globally known companies across broad range of industries 233; Gothard Base Tunnel 209–10; Human Brain Project 210; humanitarian tradition 228; immigration and 226–30; innovation rankings *12*, 203–6; international organizations, global capital for 228–9; Nobel Prizes per capita 215; Orson Welles caricature of 206, *207*; Paul Scherrer Institute (PSI) 214–15; pharma industry 218–20, 228, 233; politics in 221; quality of life in 221; recycling of talent in 222–3; scientific institutes 208, 211–15; size of 205–6; transport network/railways 208–11; universities 208, 215–21, 223–5; watchmaking 226–8, 233, 234; WIPO Global Innovation Index *12*, 203
Syngenta 23
Synthe 220, 231

Taiwan 8, 141, 187, 189, 306
Talbot, David 58
Tan, Anthony 182
Tang Xiao'ou 33–4
tax 50, 145, 152, 153, 181, 247
TD Bank 281
Tech City 105
Tech Market Cap ($T) per region (2021–2023), Total *11*
Tech Nation 128
Technical University of Munich (TUM) 240
Technopreneurship Innovation Fund (TIF) 179
Temasek Holdings 176, 183, 184, 187, 193
Temu 45, 306–7
Tencent 3, 17–23, 25, 27, 28–9, 38, 48, 51, 97, 159, 170–71, 172, 174, 178, 180, 190, 193, 241; Tencent Music Entertainment (TME) 21; Tencent Pictures 21; Wallerstein and 16–20, 22–3, 27, 29

Index

Tengger Desert Solar Park, China 205–6
Terman, Fred 76, 82–8, 94
Tesla 8, 22, 27, 28, 84, 95, 116, 146, 240, 265, 276
Texas Instruments 93
Thailand 172, 175, 186, 188, 190
Theranos 95, 134
Thiel, Peter 60
Thoma Bravo 127
Thompson, Derek 102
Thornhill, John 124
Thousand Talents Program (TTP) 73
TikTok 8, 33, 172, 190, 191
Tirole, Jean 13
Titanic 83
Toon, Nigel 129
Tortoise Responsible AI Forum 136
Toshiba 93
Toynbee, Arnold: *A Study of History* 75
traitorous eight 88
transistors 88, 128–9
Transpacific Railroad 80
transport networks 43, 80, 107–9, 208–11
Troncoso, Carmela 223
Truman, Harry 86
Trump, Donald 294, 295
Tsinghua College 33, 35–8, 40
TSMC 8, 147, 187
TwelveLabs 163, 164
Twitter 98, 310

Uber 19, 72, 98, 117, 134, 140, 161, 183, 215
Ubisoft 21, 25
UCL 108, 109; Centre for Artificial Intelligence 119–20
UiPath 125, 263–4, 306, 313
UK 46, 98–137, 140, 182, 186, 204, 209, 215–16, 220, 233, 243, 245, 264, 265, 275, 284, 292, 298, 300, 306; Advanced Research and Invention Agency, or ARIA 120; AGI, scientific approach to building 119–20; Arm sale and 123–8; Brexit 113, 114, 115, 126; colocation 110–12; DeepMind *see* DeepMind; Department of Science, Innovation and Technology 119; growth of medium-sized companies and sale to bigger companies abroad 122–3; Isambard AI 130; Matthew effect and 111–12; multidisciplinary nature of 110–11; Graphcore and problem of raising money 128–31; regional disparities in 112–18;

Science and Technology Framework 119; supporting role to US 122–8; top talent and 103, 109; top ten countries by number of private billion-dollar tech companies and *9*; transport links 108–9; trillion-dollar company as desirable policy outcome 121–2; Valley model and 131–4; venture capital and *10*, 99, 101, 103, 107–8, 112, 120, 125, 131, 132, 133; WIPO Global Innovation Index and *12*
UKRI 130
United Nations (UN) 12, 136, 143, 221, 229
universities 208, 215–21, 225; Canada 4, 31, 270, 275, 277–8, 280–6, 290, 299, 301, 302; China 31, 33, 34, 35, 36, 38, 39, 40–41; Germany 240, 259–60; Singapore 180–82; South Korea 159; Switzerland 208, 215–18, 223, 224–5; UK 103, 108, 117–18, 119; US 72, 73, 76, 79–83, 86, 87, 89–90. *See also individual university name*
University of Alberta 270, 286, 288, 289
University of Berkeley 17, 40, 90, 92, 164, 167
University of California San Diego (UCSD) 275
University of Manchester 117–18
University of Montreal 270, 282–3
University of Science and Technology (USTC) 34
University of Toronto 4, 31, 270, 275, 277–8, 280, 299, 301, 302
UnternehmerTUM 259–60
Upton, Eben 112, 122
USA 56–97; adaption to change in 92–3 AI and 10, 38, 54, 66–77, 100, 102, 109, 123, 125, 126, 127, 129, 130–31, 146, 188, 198, 245, 259, 261, 262, 272, 276, 278, 284, 289, 290, 292, 302, 310; ASEAN and 186, 187, 189, 191; Canada 4, 280–81, 283, 289, 290, 292, 295, 297, 298–300; China and 7–8, 13, 20, 22, 23–9, 32, 33, 35–42, 44, 45, 49, 51, 53, 54, 56, 67–74, 189–90, 191, 192, 304–7, 314; Chinese acquisitions of US firms, curbs 23–9; chips, restricts access to high-performance 73–4, 131; concentration of economic power in 248; Germany and 239, 240, 243–9, 251–7, 260, 262–6, 275; immigration and 226, 293, 295, 297, 298–300; rebalance or pivot to Asia policy 189; Silicon Valley *see* Silicon Valley;

USA (cont'd)
 Singapore and 170, 171, 173, 181, 182, 185, 186, 192, 199; South Korea and 139, 140, 141, 143, 146, 149, 153–4, 159, 163–4, 165–8; supremacy in new technologies, possible waning of 7–8, 304–7, 314, 315; Switzerland and 205, 209, 212, 214, 215–16, 218, 220, 222, 229, 231, 232, 233, 234, 236, 237; talent acquisition 70–73; UK and 103, 104, 106, 108, 113, 114, 119, 121, 122, 124, 126–8, 130, 131–4; universities 72–3, 76–90, 215, 216, 217, 220, 275; venture capital in 10, 24, 54, 59, 64, 77, 89, 93–7, 166, 262
U-shaped returns 94–5
UUNET 97
UVC 262

V2 rocket 239
Valium 219–20, 226
Varza, Roxanne 6
Vector Institute for Artificial Intelligence, Toronto 274, 279–84, 286, 297, 301, 302
venture capital: Canada and 279, 300–301, 309; China and 22, 23–7, 38, 42, 52; Germany and 245, 252, 254, 257, 259–64; Singapore and 170, 173, 179–82; South Korea and 157–8, 166; Switzerland and 204, 218, 231, 233 236, 237; top ten countries by volume of venture dollars invested 10; UK and 99, 101, 103, 107, 108, 112, 120, 125, 131, 132–3; US and 10, 24, 54, 59, 64, 77, 89, 93–7, 166, 262
Vernon, Raymond 91
Viazovska, Maryna 225
Vietnam 69, 147, 172, 185, 186, 187, 188, 190, 191
VinBrain 188
VinFast 187, 188
Virtual Singapore 196
Visit Sweden 205
Vitana V 154
VMware 69, 222
Vollmann, Christian 243–4
Volvo 23, 267, 316, 317
von der Leyen, Ursula 26
Voyager 1 205

Wafios AG 249, 252
Wallerstein, David 16–20, 22–3, 27, 29
Wandelbots 268–9
Wang Jian 33
Warner, Marc 108–9
WaveOptics 127
Waymo 46
wealth; distribution 53; inequality 53, 112, 254–6, 255
Web3 movement 117
WeChat 8, 18–19, 20, 21, 159
Weedbrook, Christian 292
Wellcome Trust 106
Welles, Orson 206, 207
Wiegand, Daniel 240, 241, 242
Wilhelm I, Kaiser 242
Woergoetter, Andreas 258
Wooldridge, Adrian 254; *Capitalism in America* 304
World Bank 177–8, 181
World Intellectual Property Organization (WIPO) Global Innovation Index 11–12, 12, 203–4
World Wide Web 213–14
WorldFirst 25
Wyss, Hansjörg 230–32

Xanadu 5, 292
Xi Jinping 36, 44, 53, 54, 192, 295
Xiangyu Zhang 31, 37
Xiaodong, Li 'Forrest' 169–75
Xiong'an, China 43–4
Xolo3d 261
XRD 117
Xu, Chris 191

Y Combinator 66–7, 116, 165
Ya Qin Zhang 34, 36–7, 45, 67
Yahoo 72
Yi-Large 54–5
Yin Qi 33
Ying Ma 33
Yip, Leo 198–9
YY 191

Z3 243
Zhang Taisu 53
Zhang Yiming 191
Zhipu 38–9
Zhongguancun, Beijing 33, 35
Zinal 232–3
Zuberi, Bilal 74–5